Cultural Ergonomics

Theory, Methods, and Applications

T0179296

Cultural Ergonomics

Theory, Methods, and Applications

Tonya L. Smith-Jackson • Marc L. Resnick
Kayenda T. Johnson

CRC Press
Taylor & Francis Group
Boca Raton London New York

CRC Press is an imprint of the
Taylor & Francis Group, an **informa** business

CRC Press
Taylor & Francis Group
6000 Broken Sound Parkway NW, Suite 300
Boca Raton, FL 33487-2742

First issued in paperback 2019

ISBN-13: 978-1-4398-1260-0 (hbk)
ISBN-13: 978-0-367-37905-6 (pbk)

Library of Congress Cataloging-in-Publication Data

Cultural ergonomics : theory, methods, and applications / editors, Tonya L.
Smith-Jackson, Marc L. Resnick, Kayenda Johnson.
       pages cm
    Includes bibliographical references and index.
    ISBN 978-1-4398-1260-0 (hardcover : alk. paper)
    1. Human engineering--Cross-cultural studies. I. Smith-Jackson, Tonya L. II. Resnick,
Marc L. III. Johnson, Kayenda.

    T59.7.C85 2014
    620.8'2--dc23   2013030833

Visit the Taylor & Francis Web site at
http://www.taylorandfrancis.com

and the CRC Press Web site at
http://www.crcpress.com

Contents

Preface

*Tonya Smith-Jackson, PhD, CP, Marc Resnick, PhD,
and Kayenda Johnson, PhD*

PURPOSE OF THIS BOOK

This book is intended to be an introduction and primer to augment teaching and application of cultural ergonomics to support empirical and translational research, design and evaluation. Examples and methodologies are provided within a human factors framework, with research to practice serving as the primary motivation. Human factors/ergonomics as a discipline is slowly integrating cultural ergonomics into efforts to explore human capabilities and limitations in the context of design and evaluation. The authors of the book chapters have significant experience in various aspects of cultural ergonomics, having been successful in applying the approach for many years.

DEFINING CULTURE AND CULTURAL ERGONOMICS

Cultural ergonomics applies what is known about cultural attributes to the design, development, and evaluation of systems. The goal of cultural ergonomics is to ensure systems and products are designed to be inclusive and equitable for the intended users. Even when products and systems are highly localized, rarely does one design for a single, monocultural population of users. Most nations have multiple cultures, and most individuals integrate multiple cultural identities. The products and systems created and used are cultural artifacts representing shared cognitions that characterize mental models resulting from interactions with physical environments. Thus, culture is embedded in every artifact and impacts the extent to which products are usable, accessible, useful, and safe. It is well known in human factors and ergonomics that products and systems deviating from users' mental models may have negative consequences for users, ranging from minor annoyance to more serious consequences such as severe injury or death.

Our working definition of culture is given by Veroff and Goldberger (1995) and was selected because it is grounded in cultural psychology, sociology, and anthropology. Additionally, the definition minimizes ethnocentric perspectives. According to Veroff and Goldberger, culture is a concept

> ... referring to a collectivity of people who share a common history, often live in a specific geographic region, speak the same or a closely related language, observe common rituals, beliefs, values, rules, and laws, and which can be distinctively identified according to cultural normative practices such as child-rearing, kinship arrangements, power arrangements, and ascribed roles that make up the fabric of how a society functions. (p. 10)

This definition is comprehensive, emphasizes the factors that socialize and embody users into specific cultures, but most important, helps to distinguish culture from other user characteristics such as experience and ability/disability. Additionally, the definition is not confounded by the inaccuracies regarding physical distinctions related to the arbitrary and socially constructed concept of "race." In this volume, we focus on the following constructions of culture across the globe, which include ethnicity, gender, religion, and socioeconomic status, among others (see Table 1).

TABLE 1
Cultural Attributes of User Populations

Cultural Attribute	Description	Example
Ethnicity[a]	Shared language, national origin, history	African American, Bengali, Chinese, Ghanaian, Indian American, Korean, Indian, Indonesian, Malay, Nigerian, Roma, Spanish, Swedish
Gender[b]	Identity and socialization as based on sexual cognition (is not always consistent with biological sex)	Asexual, female, gender-variant, male, queer, transgender
Nationality[b]	Country of origin, or in some cultures, nation of ancestor's origin	Aruba, China, Germany, Israel, Jamaica, Mexico, Saudi Arabia, Switzerland, United States
Religion[b]	Spiritual or, in some cases, philosophical world view and belief system	African diasporic, atheist, Buddhist, Christian, Hindu, Judaism, Muslim, shamanist
Generation[c]	Social group born in a common interval of years or decades marked by significant events, aka *turnings*	Baby Boomers, Generation X, Generation Y, Millennials
Educational level[b]	Level of education attained within a structured learning system or role related to education	Primary, secondary, university graduate (bachelor's degree), post-graduate (master's, doctorate) and their equivalents
Socioeconomic status[b]	Similar to social class and defined by combining income and education	Affluent middle-class, minorities, lower class, middle class, poor, rich, upper class
Regional culture[d]	Geographic area within a culture that may contain a comparatively homogeneous group of residents	Appalachian, Easterner (US), metrosexual, Northerner (US), rural, Southerner (US), urbanite, Westerner (US)

[a] Betancourt, H., and Lopez, S., 1995. *The Culture and Psychology Reader.* New York: New York University Press, pp. 87–107.

[b] Descriptions derived from APA (2006).

[c] Strauss, W., and Howe, N. (1994). *The Fourth Turning: An American Prophecy.* New York: Broadway Books.

[d] Gupta, A., and Ferguson, J. (1997). *Culture, Power, Place: Explorations in Critical Anthropology.* Durham, NC: Duke University Press.

While there are hundreds of definitions of culture, we made every attempt in this volume to place boundaries around our concept of culture and ensured the boundaries were grounded in sociology, anthropology, and cultural psychology. These boundaries are important for three reasons. First, bounding the concept of culture helps designers and researchers work within a shared knowledge space to organize empirical research and development efforts. Second, a bound definition of culture is useful for operationalization of culture as a variable or parameter for use in quantitative studies and within methods requiring analysis of culture. Third, when any design concept is concrete in the minds of practitioners and researchers, the difficulty of drawing inferences to support application is greatly reduced. Vague design concepts are more likely to be misapplied and may yield incorrect inferences. In addition to the strategic advantages of using a bound definition of culture, our definition also helps to delineate cultural ergonomics from other important design for diversity approaches, such as accessibility and universal design.

IMPORTANT FEATURES IN THIS BOOK

We have also incorporated features to demonstrate how cultural ergonomics can be applied in research and practitioner contexts. These features include selection of theories, descriptions of research designs, methods to analyze the results, case studies, and strategies used to draw inferences and conclusions. The application areas are vast, including occupational safety, global issues, emergency management, human–computer interaction, warnings and risk communications, and product design.

REFERENCES

American Psychological Association (2006). *APA Dictionary of Psychology.* Washington, DC: American Psychological Association.

Betancourt, H., and Lopez, S., 1995. The study of culture, ethnicity, and race in American psychology. In N. R. Goldberger and J. B. Veroff (Eds), *The Culture and Psychology Reader.* New York: New York University Press, pp. 87–107.

Gupta, A., and Ferguson, J. (1997). *Culture, Power, Place: Explorations in Critical Anthropology.* Durham, NC: Duke University Press.

Strauss, W., and Howe, N. (1994). *The Fourth Turning: An American Prophecy.* New York: Broadway Books.

Veroff, J., and Goldberger, N. (1995). What's in a name. In N. R. Goldberger and J. B. Veroff (Eds), *The Culture and Psychology Reader.* New York: New York University Press, pp. 3–21.

Acknowledgments

Research in the area of inclusion is fraught with challenges from many directions. In particular, the peer review process, if based on monodisciplinary reviewers who are fully grounded in monodisciplinary approaches, can be difficult. In this volume, we selected reviewers whose prior work and training demonstrated interdisciplinary engineering, human factors, ergonomics, and psychology foundations, and whose epistemologies, as reflected in prior work, seemed contextualized and inclusive. Thus, the subject matter experts who reviewed the chapters and suggested revisions were of the greatest caliber. As expected, they provided useful feedback to strengthen and improve each chapter. To these reviewers, we will be forever grateful.

The Editors

Tonya Smith-Jackson, PhD, CPE, is professor and chair in the Department of Industrial and Systems Engineering at North Carolina Agricultural and Technical State University. She was formerly a member of the faculty of the Grado Department of Industrial and Systems Engineering at Virginia Tech. Dr. Smith-Jackson's research focuses on the application of cognitive and cultural ergonomics to the design, analysis, and evaluation of systems, with a specific focus on safety and risk, systems design, work system analysis, inclusive design, human-systems integration, and mixed methods data integration and analytics. She graduated from the North Carolina School of Science and Mathematics in 1982, earned a BA in psychology from the University of North Carolina–Chapel Hill, and obtained her MS and PhD degrees from North Carolina State University in psychology/ergonomics (interdisciplinary ISE) in 1989 and 1998, respectively. She is a member of the Human Factors and Ergonomics Society, the Institute of Industrial Engineers, the American Society of Safety Engineers, and the Association for Psychological Science. She was certified by the Board of Certification in Professional Ergonomics in 2009. Dr. Smith-Jackson has worked in other arenas outside of academia, including industry (IBM, Ericsson Mobile Communications, and forensics) and government (Consumer Product Safety Commission, Army Community Services). She has also taught as an adjunct faculty member in community colleges and universities in New York, Virginia, North Carolina, Maryland, and Germany.

Marc L. Resnick, PhD, is a professor of human factors and information design in the College of Business at Bentley University. He applies principles of cognitive and behavioral science to human–technology systems at the strategy, design, and evaluation phases. He is also a frequent consultant for private enterprises, government agencies, and not-for-profit organizations in user experience, forensics, and innovation. He completed his doctoral studies in industrial and operations engineering at the University of Michigan, and spent 18 years on the faculty of industrial and systems engineering at Florida International University. He holds leadership positions in several professional societies, including the Institute of Industrial Engineers and the Human Factors and Ergonomics Society, and is an active philanthropist.

Kayenda T. Johnson, PhD, is a user-centered designer for the User Experience Practice, SRA International. She applies user-centered design methodologies and principles to computer interface design problems in order to maximize the user experience. Her work includes development of conceptual wireframes demonstrating user-focused interaction patterns for the design and redesign of both mission-critical desktop and web-based computer applications. Dr. Johnson completed her doctoral studies in industrial and systems engineering with focus on human factors engineering concentration at Virginia Tech. Her doctoral research explored the impact of culture on computer interface design. Prior to her role as a user-centered designer, she worked as a sociocultural analyst, enhancing her knowledge of both social and cultural factors that influence ways of thinking and behaving across our global society. Dr. Johnson continues to advance her understanding of culture's impact on design through her participation in international missions trips. Her broader areas of focus are human–computer interaction, interface design, cultural impacts on computer interface design, user-centered design methodology development, user research, and usability studies.

Contributing Authors

Taylor J. Anderson, MA, is a Research Associate III, SA Technologies, Inc. Mr. Anderson has experience in the design of user-centered interfaces created using the principles of Situation Awareness-Oriented Design (SAOD) and good human factors. He completed his master's degree in experimental–human factors psychology from the University of Dayton. Mr. Anderson has experience applying multiple types of interview techniques such as goal-directed task analysis (GDTA), critical decision method, knowledge audit, and cognitive task analysis. He has experience in design of systems in diverse domains such as unmanned vehicle command and control, global public health, oil well drilling, and antisubmarine warfare.

Sharnnia Artis, PhD, is the Education and Outreach Director, Center for Energy Efficient Electronics Science, an NSF-funded science and technology center at the University of California–Berkeley. Prior to her appointment at Berkeley, she was a postdoctoral research fellow at Ohio State University, where her research area focused on the application of cultural ergonomics to engineering education and understanding the needs and success factors of underrepresented populations in engineering. Dr. Artis received her BS, MS, and PhD degrees in industrial and systems engineering from Virginia Tech. She also has industry and government experience as a human factors engineer, having held positions at Chevron, DuPont, Army Research Laboratory, and Aptima.

Cheryl A. Bolstad, PhD, is a Principal Research Associate, SA Technologies. Dr. Bolstad has a PhD in psychology, specializing in cognition and human factors from North Carolina State University. She more than 20 years of experience as a human factors engineer working for the both the military and private sectors. She has worked extensively in situation awareness (SA) research, user interface design, SA measurement and team performance assessment, and training. Dr. Bolstad has authored more than 100 publications, is a member of multiple professional organizations, and is a certified professional ergonomist.

Haydee M. Cuevas, PhD, is an assistant professor in the College of Aviation's Department of Doctoral Studies at Embry-Riddle Aeronautical University (ERAU), Daytona Beach, Florida. Prior to joining ERAU, she worked for more than 7 years as a research scientist at SA Technologies, Inc. She has a PhD in applied experimental and human factors psychology and a BA in psychology, both from the University of Central Florida. Dr. Cuevas has more than 15 years of experience as a human factors researcher investigating a broad range of human performance issues in complex operational environments. Her research interests include human–automation team cognition in unmanned aerial systems and commercial human spaceflight operations.

Mica R. Endsley, PhD, is president of SA Technologies in Marietta, Georgia. She received a PhD in industrial and systems engineering from the University of Southern California. She has authored more than 200 scientific articles on situation awareness, decision making, decision support systems, and automation, and is the co-author of *Designing for Situation Awareness,* published by CRC Press in 2004.

Abeeku Essuman-Johnson, PhD, is a professor and chair of the Department of Political Science, University of Ghana, Legon, where he earned his BA, MA, and PhD in political science. He specializes in research methodology and applied political science. He has also conducted research in safety engineering, conflict and refugees in Africa, and organized labor in Africa. He has conducted research abroad at several universities including North Carolina A&T State University, North Carolina State University, Virginia Tech, and Carleton University.

Richard C. Goldsworthy, PhD, is CEO and director of Research and Development, Academic Edge, Inc., in Bloomington, Indiana. He specializes in the design, development, and evaluation of multimedia communications products, particularly in the areas of learning and instructional tools. He has several grants and contracts from the National Institutes of Health and the Centers for Disease Control in the areas of prevention and control, health beliefs, and behavioral change.

Diana Horn, PhD, is a principle researcher at Engineered Research, Inc. She has developed user-centered interfaces for the medical, travel, military, and communications industries. She is currently involved in leading design and development of mobile applications for hotel reservations and guest management. Diana completed her master's degree and doctoral degree in industrial engineering from Purdue University and has conducted research in creativity and customer relationship management.

Rashaad E. T. Jones, PhD, has his doctorate in information sciences and technology from Pennsylvania State University and a BS degree in electrical and computer engineering from Morgan State University. Dr. Jones's research experience includes cognitive modeling, artificial intelligence, and augmented cognition and human factors engineering. He has been involved in a wide range of projects that seek to develop cognitive models of situation awareness (SA) and has examined methods that leverage augmented cognition technologies to enhance SA. Dr. Jones also has more than 10 years of experience in computer science that includes programming, neural network development, intelligent agents, virtual reality, and modeling and simulation.

Shreya Kothaneth, PhD, is a human factors specialist at the K-12 division of Hobsons, Inc. She employs human factors and cognitive engineering principles to conduct user research and inform product design of various educational solutions. Dr. Kothaneth received her doctorate in industrial and systems engineering, with a specialization in human factors and ergonomics, from Virginia Tech. Her dissertation research revolved around the relationship between organizational culture, usability, and instructional technology acceptance. Her research interests include cultural ergonomics, usability, user-centered design, and applications of mixed methods for user research.

William Lee, PhD, is a staff member of the HSI, Visualization, and Decision Support Department, MITRE Corporation. Will conducts applied research analyzing and understanding human behaviors through statistical modeling. He also studies human performance in collaborative command and control environments. Will received his PhD in industrial and systems engineering from Virginia Tech, his MS in uncertainty visualization from the University of Alabama in Huntsville, and his BS in computer science from Middle Tennessee State University.

The author's affiliation with the MITRE Corporation is provided for identification purposes only and is not intended to convey or imply MITRE's concurrence with, or support for, the positions, opinions, or viewpoints expressed by the author.[*]

Christopher B. Mayhorn, PhD, is an associate professor and program coordinator of the Ergonomics/Experimental Psychology Program, North Carolina State University, having joined the faculty there in 2002. He earned a BA degree from The Citadel (1992), and an MS degree (1995), a graduate certificate in gerontology (1995), and a PhD degree (1999) from the University of Georgia. He also completed a postdoctoral fellowship at the Georgia Institute of Technology. Dr. Mayhorn's current research interests include everyday memory, decision-making, human–computer interaction, and safety and risk communication, as well as home medical device design, for older adult populations. Dr. Mayhorn has more than 30 peer-reviewed publications to his credit and his research has been funded by government agencies such as the National Science Foundation and the National Security Agency.

Brannan R. McDougald, MS, is a graduate student at North Carolina State University in the Human Factors and Ergonomics Psychology Program. Brannan earned his BS from the University of Georgia (2005) and an MA from Middle Tennessee State University (2008). Currently he is completing his final semester of classes. He has successfully proposed his dissertation investigating the effects of persuasive technology on the decision-making process. Additional professional interests include user-centered design, user research, interface design, and video games.

Michael S. Wogalter, PhD, is a professor at North Carolina State University in the Department of Psychology. He earned his BA from the University of Virginia, MA from University of South Florida, and his PhD from Rice University. Dr. Wogalter has published several hundred journal papers and conference proceedings, with a specific emphasis on warnings and risk communications, as well as usability and product design. He is a fellow of the Human Factors and Ergonomics Society and has held faculty positions at the University of Richmond and Rensselaer Polytechnic Institute.

Contributors

Yu-Hsiu Hung, PhD, is an assistant professor in the Department of Industrial Design of National Cheng Kung University (NCKU) in Taiwan, the director of the User Experience Research and Design Laboratory at NCKU, and a member of Alpha Pi Mu, Industrial Engineering Honor Society. Prior to joining NCKU, he was a research associate of the Assessment and Cognitive Ergonomics Laboratory in the Industrial and Systems Engineering Department at Virginia Tech, where he received his MS and PhD degrees with focus on human factors engineering. In addition to his academic pursuits in the field of engineering, Yu-Hsiu also received his MS degree from NCKU and his undergraduate degree in industrial design from Chang Gung University at Taiwan. His research focuses on service design, healthcare human factors, human computer interaction, and occupational safety and systems analysis. He has professional work experience at the Global Consumer Group at Whirlpool Corp., as a user experience researcher. He has also worked as an industrial designer at Image Corp., a rapid prototyping services company in Taiwan.

Manutchanok Jongprasithporn, PhD, is a postdoctoral researcher in the Grado Department of Industrial and Systems Engineering at Virginia Polytechnic Institute and State University and an instructor in the Department of Industrial Engineering at King Mongkut's Institute of Technology Ladkrabang. She attained her first MS degree in industrial engineering from Clemson University. She received her PhD and a second MS degree in industrial and systems engineering at Virginia Institute of Technology with concentration on human factors and biomechanics. Her main research interests include age-related effects on biomechanics of slips and falls, slips/falls training and safety, motor control, and kinesthetic learning.

Enid Montague, PhD, is a human factors and ergonomics engineer and assistant professor in the Division of General Internal Medicine and Geriatrics of the Feinberg School of Medicine at Northwestern University. Dr. Montague's research explores the role of trust in technology in systems. She studies why humans trust or distrust technology, the effects of trusting attitudes on system performance, and designing for appropriate trust.

Kathleen Mosier, PhD, is a professor of industrial/organizational (I/O) psychology at San Francisco State University. Dr. Mosier received her PhD in I/O psychology from University of California–Berkeley and her training in human factors at NASA Ames Research Center. She is a past president of the Human Factors and Ergonomics Society. She has been conducting research in human factors and expert decision making for more than 20 years. Her recent work examines communication and problem solving in distributed teams, as well as the impact of computers and automation on decision making in aviation.

Manuel A. Pérez-Quiñones, DSc, is the associate head for Graduate Studies and associate professor of Computer Science at the Virginia Tech. He is also a member of the Center for Human–Computer Interaction. Pérez-Quiñones holds a DSc from George Washington University, Washington, DC. His research interests include human–computer interaction, personal information management, user interface software, and educational/cultural issues in computing. He has published over 75 refereed articles, coauthored 10 book chapters, and is currently working on a book on user interface software. He is a member of the editorial board for ACM's Transactions on Computing Education. He was also chair of the Coalition to Diversify Computing (2010–2011), a CRA/ACM/IEEE-CS committee. At Virginia Tech, he has served as chair of the Hispanic/Latino Faculty and Staff Caucus, associate dean and director of the Office for Graduate Recruiting and Diversity Initiatives of the Graduate School, and a Multicultural Fellow.

Ryan L. Urquhart, PhD, is a senior user experience designer at Blue Cross Blue Shield of North Carolina (BCBSNC). Prior to joining BCBSNC, he was an advisory software engineer at International Business Machines (IBM), located in Research Triangle Park, North Carolina, where he provided human factors support for IBM's WebSphere and Tivoli software brands. Dr. Urquhart obtained his doctoral degree from Virginia Polytechnic Institute and State University in industrial and systems engineering, with a concentration in human factors engineering. He received his master's and bachelor's degrees in industrial and systems engineering from North Carolina Agricultural and Technical State University. His expertise is in the area of human audition, noise and performance, speech intelligibility, and human–system interaction. Over the span of 10 years, he has worked with clients such as Royal Bank of Scotland, Thomson Reuters, Ericsson, Adaptive Technologies, Inc., LG Electronics, and Army Research Laboratories.

Woodrow W. Winchester III, PhD, is an assistant professor of industrial and systems engineering in Virginia Techn's Grado Department of Industrial and Systems Engineering. He is also an affiliate faculty member for Virginia Tech's Center for Human–Computer Interaction. Dr. Winchester earned his PhD in industrial and systems engineering from North Carolina Agricultural and Technical State University, Greensboro. His area of expertise is human factors opportunities in the design of health and wellness technologies.

1 Cultural Ergonomics
Overview and Methodologies

Tonya Smith-Jackson and
Abeeku Essuman-Johnson

CONTENTS

CULTURAL ERGONOMICS JUSTIFIED

Since its inception as a formal approach by Alphonse Chapanis (1974), cultural ergonomics has gained prominence in efforts to globalize science and engineering. Cultural ergonomics is a necessary component to research, design, and evaluate inclusive systems and technologies in cross-national, global, and even localized contexts. Although some have embraced this approach, the status quo of human factors/ ergonomics continues to give priority to designing for "normal operators" working under "customary conditions" (Kroemer, 2006, p. 3). Yet, the practical concepts of *normal* and *customary* represent an ongoing bias toward designing for adults who were formerly employed in offices or manufacturing settings, ranging in age from 20 to 50, and living in Westernized affluent countries such as North America, Europe, New Zealand, and Australia (Kroemer, 2006). Even within these nations, the adults who are the focus of design are more than likely to be of European descent, excluding minority, indigenous, or aboriginal populations in those same nations. This highly constrained view of human factors/ergonomics continues to threaten the pace in which we advance science and engineering, and undermines our ability to conduct effective, usable, and safe translational research and technology transfer. Aaron Marcus (2006) suggests it is not only people in the United States who do not understand the importance of cultural differences in how we design, but also those in all

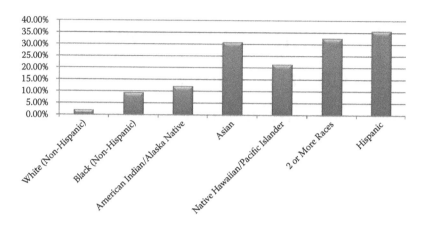

FIGURE 1.1 Percent change in population from 2000 to 2009 based on U.S. Census Data (2010).

other nations who design artifacts reflecting signs, values, and rituals that impact preferences and acceptability of the products.

The need for cultural ergonomics is clear if we simply examine human variation across the globe. The world's population hit seven billion in 2011 (Population Reference Bureau, 2011), and will exceed nine billion by 2050, with more than 80% of the growth occurring in less-affluent nations such as sub-Saharan Africa, India, and Latin America (United Nations Population Fund, 2009). Currently and through 2100, Asia and Africa will continue to be the most populous areas, followed by Latin America and the Caribbean (United Nations Population Fund, 2011). Because of higher rates of growth among minority groups (Figure 1.1), the U.S. Census projects a majority–minority crossover in 2042 resulting in a national population that is predominantly non-White, collectively older, urban, and with more economic power and educational advantages among women (U.S. Census data 2010; Johnson and Kasarda, 2010).

Other nations are also experiencing demographic changes that introduce imperatives for changing how we think about technology, design, and research on user requirements. Besides current and future population growth among cultures that are not traditionally considered in user-centric design, other global requirements that continue to underscore the urgency of expanding our use of cultural ergonomics approaches include:

- Human migration to larger metropolitan areas, introducing more diverse workers to traditional, formalized work systems (e.g., manufacturing, services, education, healthcare, transportation, public safety, tourism)
- Increasing dominance of "informal" work systems across the globe reflecting the need to systematically investigate these informal, unregulated work systems, the technologies they use, and work practices (e.g., street vendors, street hawkers, noncertified/non-licensed doula systems and midwifery, unregulated contracting in construction, seasonal and migrant work, day laborers, housekeepers)

- Rising affluence of previously less-affluent population sectors leading to longer life spans, better access to quality healthcare, and more access to consumer products and services, recreation, education, and transportation systems
- Global climate change, humanitarian challenges, conflicts, and natural and human-caused disasters that require cross-national cooperation and public engagement
- Secular trends within populations due to access to better nutrition and healthcare, leading to more intravariation in body size, body proportions, earlier onsets of puberty, and increasing weight

One important manifestation of the evolution of our increasing understanding of the importance of diversity is the recent focus on informal work systems as a critical area of study to enhance the safety, efficiency, and effectiveness of the workplace. Scott (2009) and others have led the field in task-analyzing many forms of work in countries around the world, and have brought attention to the informality of work systems when compared to the formalized, explicit work systems in the more affluent countries. The informality of work in industrial developing countries is the predominant mode of operation, where work is not written down, trained in, or analyzed to support further review and redesign. In most counties, work is simply done in the manner so chosen by hired hands, with equipment that is available, some of which might be homemade, and at times that fit whomever has the funds to pay laborers.

Specifically, informal work systems are critical for three reasons. These forms of work organization, although comprising 8.4% of the US Gross Domestic Product (GDP), actually comprise from 40% to almost 70% of the GDP of countries in Asia, Latin America, Africa, and Central and Eastern Europe (Schneider, 2006 in Jutting, Parlevliet, and Xenogiani, 2008). Not only are workers using technologies and work practices within informal work systems, but they are also producing and delivering goods in these same systems without the respective protections from injuries and fatalities, assurances of usability, or training in work practices to enhance efficiency and safety. Even formal work systems must still struggle with designing work practices and technologies for an increasingly diverse workforce, in spite of data and continued research in formal work. Thus, the comparatively less-studied informal work systems are in drastic need of the same level of research and design scrutiny afforded to more formal work systems.

Service systems such as healthcare are also impacted by cultural diversity, at least in terms of the quality and safety of services that can be delivered. Health and wellness technologies such as wheelchairs, lifts, toilets, shower chairs, hospital beds and showers, imaging machines (i.e., fMRI, CAT), and blood pressure cuffs may not be adequately designed to support the delivery of quality healthcare and patient self-management. The implications of poor design in the aforementioned technologies could be significant and may be driven by something as simple as anthropomorphic variation. ISO/TR 7250-2:2010 is a useful standard to support inclusive design based on physical differences across nationalities (ISO, 2010). These trends will alter the landscape of users of technologies for health and wellness and make paramount the importance of educating and training to increase the cultural competence of healthcare providers, as well as human factors researchers and practitioners.

Nations with high adoption rates for technologies such as information and communication technologies (ICTs) have clear evidence from the past three decades of the impact of technology on economic and individual advancement within the culture. There are, however, several nations with low adoption rates of, for example, ICTs. In Ghana, West Africa, societal factors such as policy and the quality of the infrastructure present new design challenges that should be considered when designing at the micro level. The reliability of the support structures around a technology will impact how and how well the technology can be utilized (Frempong, 2011), yet designers often overlook the context of deployment of technologies. Cultural competence in design would also consider the advantages of launching computer applications on mobile technologies rather than on desktops and laptops. Data from Frempong (2011) indicated that mobile phone proliferation in Ghana was much higher compared to televisions, radios, and land lines. These considerations introduce diversity to the traditional views of design and evaluation, but also remind us of the importance of thinking at the macroergonomic level when developing products that will be used across cultures.

ADVANCES IN THE KNOWLEDGE DOMAIN

There are several important advances in the knowledge domain that have contributed to the progress of cultural ergonomics, yet we are unable to identify an exhaustive list because communications across the globe and access to information on an international level is still fraught with major limitations or restrictions. Only recently have open access journals become more prominent and considered prestigious, resulting in more knowledge sharing among researchers in cross-cultural specialties. And, in the past two decades, agencies such as the National Science Foundation, have allocated more funding to support international collaborations, when previously, such agencies as Fulbright and the United States Agency for International Development (USAID), were the sole providers of limited research funding for international collaborations. Likewise, many professional societies have recognized the importance of international members and some are holding conferences outside of the United States and Europe, albeit somewhat financially out of the research of many. Two societies demonstrating major commitments to cultural ergonomics are Human Factors in Organizational Design and Management (ODAM) and the Institute of Industrial Engineers. In 2011, ODAM held its 10th International Symposium in Grahamstown, South Africa—the first time a symposium was held on the continent of Africa by any of the professional societies that are part of the International Ergonomics Association. Another example of commitment to cultural ergonomics is shown in IIE, which dedicates several paper sessions to research in industrial and systems engineering in Latin America during its annual Industrial and Systems Engineering Research (ISERC) Conference. The papers and presentations are in Spanish and enjoy very active participation, while also producing venues for collaboration and advancing intellectual products in industrial and systems engineering. Although the response has been slow, professional societies are beginning to recognize one major principle of scientific advancement: to be effective, any intellectual endeavor must be relevant and accessible. Relevance in the new millennium is globally motivated. But, most important, the expansion of global communications about science, technology, engineering, and mathematics (STEM) has helped to advance principal theories and approaches that enabled such specialties as cultural ergonomics.

One such advancement has occurred in how science is perceived and understood. Pejorative dichotomies such as "hard" and "soft" science are no longer considered appropriate categorizations, and such ideas as "pure" science have fallen by the wayside as inaccurate descriptions of a phenomenon that is now understood to be biased and originating from narrow Western-centric worldviews. In spite of claims of objectivity, scientists cannot separate their own mental models (derived from social and cultural contexts) from how they approach scientific research, analyses, inferences, and applications. These mental models are taught by and institutionalized within a rigid system of education and training, and within an implicit and vicarious reinforcement schedule that rewards rigid adherence, and punishes deviations or challenges. Harding (1993) summarizes the views of several historians of science that have discussed the inability of science to free itself from "racist and Eurocentric assumptions" (p. 11) and who have made claims that "sciences will be no more emancipatory than are the larger social agendas that nourish and guide knowledge seeking" (p. 11). The social agendas continue to be dominated by policymakers and research funding that is limited by exclusionary worldviews that support certain epistemologies or ways of knowing, and in particular, exclude certain culturally derived frameworks for understanding the natural world and humans who interact within the natural world. In combination, the systems of peer review slow the progress of innovation, since peer reviewers must serve as agents or gateways to access through enforcement of guidelines and criteria set forth by the same exclusive agendas. On the one hand, the culture of the scientific enterprise is essential to ensure quality in research and development. However, quality is not ensured by exclusive adherence to ethnocentric ways of knowing, but through the recognition of innovative mental models of science that are relevant to diverse cultural contexts, awareness of the politics and biases influencing the culture of science, and awareness of the complexities of the human experience. In this volume, we expand on the methodological implications of an expanded, inclusive mental model of scientific research.

As stated, there are numerous advances in the knowledge domain that have made way for cultural ergonomics as a justifiable and rational approach to research and development, but the advances are too numerous to discuss in one volume. There are, however, a few additional theories that also help to organize our thinking around the application of culture to the design and evaluation of products and systems. These include (1) the TRIOS framework, (2) the Cultural Lens Model (CLM), and (3) embodied cognition.

THE TRIOS FRAMEWORK

TRIOS is a metaframework that seems to be a practical means to organize several theories in cultural psychology, sociology, and anthropology. TRIOS is an acronym for time, rhythm, improvisation, oral expression, and spirituality (Jones, 1979; Figure 1.2). The framework arose from cultural psychology and is used to understand behaviors, attitudes, and beliefs within a variety of contexts. From cross-cultural

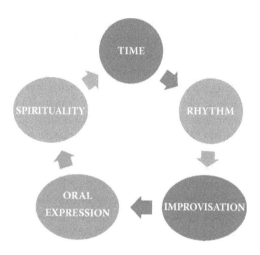

FIGURE 1.2 Dimensions of the TRIOS model. (From Jones, J. M. (1979). *Research Directions in Black Psychology*, pp. 390–432. New York: Russell Sage Foundation.)

research, the TRIOS categories provide a basis to examine cultural differences to support design theory and application. For example, consider the TRIOS framework when applied to learning and training models. Research by Waddell (2010) and Willis (1989) reveals the incompatibility between traditional design of pedagogy and learning environments and the actual learning styles and preferences of African Americans. In spite of applications of Vygotsky's theories that emphasize social interaction as a key element of learning, the degree, manner, and intent of social interactions to facilitate learning varies in preference between cultures. Among African-American learning styles and consistent with the improvisation and oral expression dimensions of the TRIOS framework, learning in interdependent social groups and allowing creative, innovative, and variable ways to apply and discuss concepts holds strong precedence over other ways of learning. If one examines the current learning environments for the education and training of adults and children, classrooms with desks in rows and columns, "death by PowerPoint" (i.e., mono-directional. and nonvariable information delivery), restriction of application of concepts, directed problem solving, and limited opportunities for oral expression are still tolerated as acceptable teaching–learning methods. Unfortunately, as urban public institutions focus more on behavioral control, the use of TRIOS-based teaching–learning is less likely. Ultimately, the efforts to control behavior may lead to unintended consequences such as suppression of learning, further disengagement from the knowledge domain, and a separation or lack of connection to other learners and learning communities.

But the question remains as to the relevance of TRIOS to design. When integrated with Hall's (1966, 1983) dimensions, the implications for design can be clarified. Like Hall's dimensions, TRIOS is based in social anthropology, and essentially reflects the social cognitions arising from ecological interactions. As an example, the improvisation dimension of TRIOS overlaps with the dimension of risk in Hall's list of dimensions. Specifically, some cultures are more comfortable with risk and uncertainty while others are less comfortable. This comfort plays out in the need for information, tolerance for ambiguity, and flexibility in operations and processes. Similarly, uncertainty avoidance in Hofstede's (1997) model describes the preference patterns with one condition: improvisation and uncertainty avoidance are direct opposites. Strong preferences for improvisation are similar to low levels of uncertainty avoidance.

For efforts toward internationalization of Web-based and mobile applications, several researchers have manipulated the degree of structure, menu depth, predictive displays, and other design features to produce flexible and broad designs to facilitate global adoption (Jhangiani, 2006; Jhangiani and Smith-Jackson, 2007; Zahedi, 2001). Consistent with Barber and Badre (1998), like any product, interfaces have cultural markers that serve as features reflecting the cultural customs and preferences of users. The use of multiple means to connect socially within one mobile product would have high appeal to users with collectivist values, and within the TRIOS model, would support certain cultures' recognition of the value of interdependence. Ferreira (2002) compared collectivism and interface design elements for an e-commerce application among Anglo-American users and Hispanic American users whose profiles indicated high or low acculturation, where high acculturation reflected strong identification with Anglo-American culture. The experimental design manipulated interface type (collectivist or individualist) and product type. Fifty-one percent of Anglo-Americans and highly acculturated Hispanic Americans chose collectivist interfaces, while 68% of low-acculturated Hispanics chose collectivist interfaces. Of note is the predominant choice of collectivist preference across all three user groups. Although the differences were not statistically significant, the results were interesting in terms of the overall preference for collectivist interface elements within this sample of college students.

CULTURAL LENS MODEL

Some frameworks have emerged inductively from military research, such as the CLM. Helen Klein (2004) provides a framework that focuses on the cognitive meta-schemas that differ between international groups. Her framework is based on years of research and observations of different cultural groups during cross-cultural collaborations. As an example, she describes the potential for conflict when different cultural groups with different worldviews collaborate:

> When people differ in the cognition and the behavior/social dimensions important for a particular natural setting and task, there is a potential source of conflict and failure. People cannot adjust mismatches by altering their underlying cognitive processes, that is, how they think about the world. Differences on dimensions cannot be changed at will because they reflect the demands of earlier experience. Even when the dimensions cause conflict in a new ecological and social context, they tend to persist. Although people can learn new content, it is difficult to acquire new reasoning forms. (p. 11)

According to Klein, the mismatches in worldviews occur because of differences in origins (i.e., cultural identity, life experiences) and differences in cognitions (e.g.,

FIGURE 1.3 Illustration of posited developmental stages to achieve self-globalization. (Smith-Jackson, Brunette, Artis, Johnson, Perez, and Resnick, 2008.)

schemas, mental models, language, affect, perceptual styles). If adjustments are not made, the resulting collaborations will not be effective and the resulting designs may fail as well. Users and designers may also originate from different cultural groups. Key to Klein's model is the idea that neither the user nor the designer/researcher can alter underlying cognitive processes and patterns easily. Yet there are many organizations, companies, and researchers who believe designers can be easily trained to think like users even when the users represent a different culture from the designer. This persistent belief is fundamentally flawed from both a psychological and engineering perspective. The development of the level of cultural competence needed to adjust one's cognitive patterns goes well beyond a company training session. Cultural competence emerges from a combination of interpersonal and intrapersonal intelligences that support such cognitive acts as perspective taking, recognition of cultural cues in verbal and nonverbal expression, knowledge of cultures and people, lived experience within diverse ecologies, and an ability to recognize, adjust, and adapt dynamically and quickly when necessary. This set of capabilities allows a researcher or designer to adjust their cognitive processes so they can better encode and process critical design data and information as it is conveyed by users. This process has been referred to as "self-globalization" (Smith-Jackson; Figure 1.3) reminiscent of Bandura's self-actualization theory (1977, 1982.). Self-globalization is an attitude that influences how the researcher and designer approaches problem definitions and the subsequent research and design efforts that follow. Kroemer (2006) and Ossewaarde (2007) discussed the need for more culturally competent and self-globalized researchers and designers and the need to avoid, whether purposefully or unconsciously, mental models that yield the design of systems and products geared toward an imaginary "normal user" more likely to represent Western cultures. Researchers and designers are challenged to expand their knowledge base and worldviews, while also seeking opportunities to increase cultural competence through education, but even this level of advancement will not make application of cultural ergonomics simple or straightforward.

As with any cultural framework, the challenge to apply the results of applied research on culture and design is significant. Several researchers offer systematic methods to translate cultural information to design guidelines and/or user requirements (Alostath, Almoumen, and Alostath, 2009; Marcus, 2006; Smith-Jackson, Nussbaum, and Mooney, 2003; Weber and Hsee, 1999). It is important to use a systematic, structured process that can be validated for design and usability in a manner that includes the target group and not simply the designers themselves (or their representatives). In a study comparing Indian and German users' product-use behaviors, Honold (2000) provided an overview to extract requirements from qualitative data derived from observations and interviews, and that data can be validated by a new set of users. Similarly, Harel and Prabhu (1999) introduced the Global User Experience (GLUE) method as a process to identify and validate cultural characteristics that transfer to product design. Harel and Prabhu defined *cultural characteristics* as "core cultural values that individuals in a society may share regardless of their unique perspectives" (p. 206). GLUE is used by Kodak to support design decisions, theory, and requirements for its global products. In the first step, ethnographic research is used to identify *product appearance qualities* that are preferred by target users. This step is used to develop 3D models that are then used in the second step, referred to as validation. The validation step includes focus groups and interviews of target users to determine the cultural validity and contextual relevance of the 3D models. The validation step in this process is crucial because it affords an opportunity for users to communicate to designers the extent to which the product qualities in the 3D models are indeed culturally compatible and usable rather than exaggerated, irrelevant, or worse, offensive.

Further complexities are introduced by the differences in how we interact with or engage with products and systems; in other words, the level of embodiment. In parallel with global demographic and economic requirements, our understanding of the foundations of engineering design and applied science have evolved from the disembodied abstractions that served only to limit innovation and advances in usability and safety to a better awareness of the biopsychosocial and cognitive capabilities of humans, informed by human ecologies. Instead, the foundations that drive how we approach designing and evaluating simple and complex products and systems for multiple users have slowly begun to manifest the inevitable mental models and knowledge associated with the safe and efficient use of products and systems. These are embodied and reflective of the shared and personal ecologies of the intended users. Personal ecologies are only one aspect of the cultural metaschema, but an important one. Lakoff (1987) used the term "conceptual embodiment"; the theory of how mental models and categorizations are "a consequence of the nature of human biological capacities and of the experience of functioning in the physical and social environment" (p. 12). Not only are mental models developed through interactions with the physical world, but the overwhelming majority of our interactions with

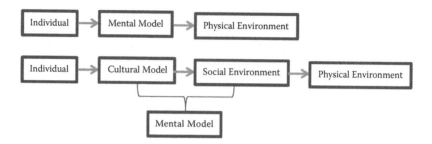

FIGURE 1.4   Cold cognitive models (upper) contrasted with a sociocognitive model (lower) emphasizing the role of culture and social environments in the acquisition and organization of mental models. (Adapted from Shore, 1996.)

products and systems are functionally embodied, in that they are mostly implemented through nonconscious processes rather than conscious thought (Lakoff, 1987). Additionally, reason and emotion are no longer categorized as polar opposites, but as synergistic processes that enable effective cognition and decision making as we engage with products and systems. But what results from the new complexities of such realizations as the embodiment of technologies is a more open dialogue about shifts in focus from older, more traditional cognitivist approaches to socioecological and social cognitive models.

In the present day, human factors and ergonomics as disciplines are no longer dominated by cold cognitive models or models that rest on abstract representations as the basis of information processing. Indeed, such areas as social cognition, hot cognition, motivated reasoning, gesture-driven memory cues, affective cognition, and implicit associations have changed the landscape and knowledge domain. Paradigms have been expanded to reflect intellectual advances in how the mind operates to support users' interactions.

These advances have important implications when designing for users whose personal ecologies and cultures differ because of differences in interactions with a variety of physical environments. These embodied interactions are used to formulate mental models of interactions with physical objects and become elements of larger schemas (Smith-Jackson, Iridiastadi, and Oh, 2011). Culture is simply a metaschema or worldview, and becomes the prime or cue for the retrieval of mental models in everyday interactions with the physical environment (Shore, 1997; Smith-Jackson et al., 2011). Shore contrasts previous cognitive models that have been replaced by sociocognitive models that recognize the influence of social environment on the individual's interaction with the physical environment (Figure 1.4). Mental models are organized through interactions with cultural and social environments. Without education and training in these topics, the best design intentions may well be lost in translation especially when users originate from different cultures as illustrated in Figure 1.5. In Figure 1.5, many users would be stumped by the prohibition against smoking while walking, and the graphic is equally confusing. More education and training on design theory within the engineering context would likely improve the outcomes of design efforts. However, effective education and training rests on a

FIGURE 1.5 A sign posted in a public area. Designers likely had the best intentions to protect public health by restricting smoking, but the poor translation undermines most readers' ability to comprehend.

shared understanding of the foundations and practices of cultural ergonomics, some of which are summarized in the next sections.

CULTURAL ERGONOMICS RESEARCH METHODS

INCLUSIVE RESEARCH AND DESIGN

The terms *inclusive research* and *inclusive design* are often met with defensiveness when raised as a challenge or need within the research community. Nonetheless, the zeitgeist has moved in a direction that requires a diverse, cross-cultural, inter-disciplinary, and broader view of design practice. In this volume, we have gently co-opted the use of "inclusion" to avoid the almost exclusive focus on individuals with disabilities among some researchers. Inclusion in the activities of research and design represents an awareness of the appropriate research approaches, recruitment, methods, analysis techniques, and interpretive frameworks that yield outputs that have equitable benefits to diverse target users.

It is less common for researchers and designers to focus solely on a user group that is exactly alike from a cultural perspective. Homogeneity is, more often than not, a fallacy within any sample of users. Most products and systems will be used in multicultural contexts, thus the idea of using the results of one cultural group to generalize to multiple users from other cultures is flawed and, in fact, could be very dangerous depending on the intended outcomes of the study. Peters and Peters (2011) discussed elevated risks to patients when determining drug dosages, primarily due to the aggregation and erroneous assumptions extended from controlled studies that failed to consider the importance of human variation. According to these researchers "even the most professionally trained among us seem to think and act as if people could be described with one simple metric or categorization" (p. 32). An aggregate used from the results of a tolerance study might set a standard dose that fits only those most represented in the actual study, while others who deviate from the participant sample may have higher or lower tolerances for the standard dose (Peters and Peters, 2011). Essentially, inclusive research and inclusive design address the shortfalls of our "traditional" approaches to conducting research and provide a framework to question traditional assumptions. Consequently, researchers should achieve higher-quality research and science that yield more equitable and usable outputs for all intended users.

Science and the practices endorsed as science such as the scientific method represent a specific way of acquiring, valuing, and internalizing knowledge. Science as a process provides a means to capture, verify, and disseminate knowledge to be used by relevant others. However, what constitutes the scientific method often includes positivist and reductionist forms of reasoning and analysis that may not always provide the most valid means to identify and understand phenomena of interest. As positivism dictates, only that which can be sensed and quantified should "count" as knowledge. To meet this mandate, researchers used reductionism or the breakdown of phenomena into the smallest, measurable forms and developed research methods to meet these constraints. These research methods emphasized internal validity by controlling experimental settings, variables, and exposures. Interestingly, even participants' responses (subjective reports) were often controlled or limited by the dependent variables researchers chose to measure. The dependent variables of choice may not have included the full response set comprising the natural reactions of participants to the environment. These are the traditions of research that extend from a sometimes unshakeable commitment to positivist, postpositivist, and reductionist worldviews.

We cannot blame researchers for being protective of the traditions under which they have been educated and trained. Some rarely deviate from their learned philosophies of how to explore problems in the natural world, frame research questions, or analyze data. A deviation from these traditions symbolizes a rejection of a knowledge-based culture, but additionally, represents a deviation from the prevailing and most popular research or design practices. This resistance to change or novelty is an advantage in some ways; it ensures we perpetuate the good practices of our disciplines and interdisciplines, and provides researchers with certain levels of success in peer-reviewed competitive arenas. But resistance to change in the face of logic also presents some disadvantages, because this resistance leads to a failure to modify methods and research approaches to enhance the quality of research over time to address new global complexities.

While describing Microsoft's use of anthropologists and ethnographers in design, Marcus (2006) reflected on the preenlightenment period in design where "anthropologists crept nervously and invisibly around the edges of CHI, UPA, HFES, and other usability and user-interface design professional conferences, seemingly hesitant to announce themselves and uncertain of the importance in which their profession was held" (p. 62). Perhaps the wait was not in vain; due to a number of design disasters and faux pas, culturally informed user experience research is finally gaining ground.

Additionally, opportunities are missed to improve the rigor of research and design practice or expand our lenses to yield higher-quality, more valid, or truthful research. Studies that deviate from the traditionalists' beliefs may also be rejected outright for publication and fail to compete in the race for funding that often characterizes the research enterprise. Interestingly, the use of the label "rigor" to describe research rarely includes a real understanding of the association between sample attributes, external validity, and ecological validity. Usually, references to rigor emphasize specific study design factors (internal validity) such as experimental design, and the quantification of every possible factor in the study. But, these forms of linguistic framing in the scientific arena are hyperfocused on internal validity at the expense or oversight of other types of rigor, that is, sample diversity, sophistication of comparative analyses, reliability of the data, sensitivity of qualitative indicators, and usefulness in the targeted context.

Resistance to such ideas as inclusive research or inclusive design is also reinforced by an implicit form of confirmation bias that is prevalent in research today, especially from the perspective of researchers with limited cross-cultural competence. In much of scientific research, the majority of researchers originate from one cultural group or very similar cultural groups, and the participants included in most studies are similar to researchers' cultural groups. As researchers verify, evaluate, and validate the outcomes of such studies, they are certainly likely to perceive the research to be accurate and rigorous because it is reflective of their own worldview and fits well within their own implicit frames of reference. This nonconscious bias perpetuates an increasingly inadequate research enterprise.

The case for nonconscious bias manifests more concretely in the "design of things." Designers hold implicit or nonconscious perspectives and biases that influence how and what they design (Akin, 1990; Barab, Dodge, Thomas, Jackson, and Tuzin, 2007; Eckersley, 1988; Greenwald, McGhee, and Schwartz, 1998). The term "frame of reference" is used in many design disciplines, including fine arts and architecture, to refer to the perspective or mental framework designers use to envision and create an artifact. Design is a creative and analytical process that is influenced by the designer's frame of reference, and designers, at times, are limited and constrained by their own frames of reference. These constraints may negatively influence creativity, compatibility, usability, and usefulness of a technology; all could be impacted negatively (Figure 1.6). Educators in design often speak of "breaking out of the frame of reference," which refers to finding experiences to assist designers in escaping their own frame of reference so they can reformulate cognitions to improve the design process and the resulting technology. Rather than basing the research solely on methodological selections (e.g., double-blind randomized control) or by the number of peer-reviewed papers or citations, quality is assessed by examining the usefulness of the research for the target users. It is difficult to be truly objective in this regard, yet this level of subjectivity is used each day to assess the quality of studies conducted under the research enterprise.

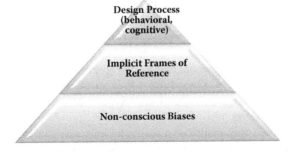

FIGURE 1.6 Iceberg model of the design bias.

Some researchers have called for a new way to assess quality of research. Rather than basing the research solely on methodological selections (e.g., double-blind randomized control) or by the number of peer-reviewed papers or citations. Washington and McLloyd (1982) suggested five key dimensions when designing and evaluating research that should be used as an a priori guide: (1) cultural, (2) interpretative, (3) population, (4) ecological, and (5) construct validities As an example, questions posed by the researcher while defining and scoping the problem might include a preliminary exploration of the extent to which cultural factors are key in the phenomenon of interest or the attributes of the population of study that introduce levels of diversity or other variances that might impact the phenomenon of interest. These five dimensions are necessary to assess the external validity of studies. Paine (undated) suggests that to be successful with inclusive research, the question of meaningfulness must be answered. How useful or meaningful was the research? To what extent did the research or design work address the problem that was initially raised as the motivation of the research? And, is the outcome or solution presented through the research meaningful, feasible, and realistic for the target group or population? Paine provides a helpful list of criteria to assess quality as follows, but a caveat is that researchers should be aware of the aforementioned assumption of heterogeneity within most populations we study:

- Used appropriate and well executed research methods
- Produced usable and accessible results
- Influenced policy and practice
- Enhanced service delivery
- Added to the evidence base
- Facilitated user involvement, as appropriate, at different stages in the research process
- Built capacity
- Represented value for money
- Developed productive relationships
- Made a difference to respondents' lives (p. 4)

It is not unusual to read published research in prestigious journals where benefits to the target population are inequitable; the benefits exist for majority group members but benefits to minority groups (including women), whose ecologies might differ dramatically, are much less apparent. With useful tools, researchers can streamline the process of designing and implementing inclusive research. Smith-Jackson and Essuman-Johnson (2011) developed a list of criteria specific to interdisciplinary engineering and design practices that reflect an inclusive approach. The list was based on several years of research, along with observations of researchers in their respective domains. The checklist (Table 1.1) is a tool to help researchers develop plans for inclusive research or design and provides a 360° evaluation tool after the work has been completed.

Efforts to increase the quality of research by enhancing inclusion and facilitating cross-cultural competence among researchers have been communicated through publications across numerous disciplines over the past decades. In response to the

TABLE 1.1
Inclusive Research Checklist

	Attribute	
1	Problem is defined by target users	_____
2	Problem definition reflects real-world needs and requirements of target users	_____
3	Research approach values subjective experiences of target users	_____
4	Research approach values objective measures to complement subjective experiences	_____
5	Methods mix qualitative and quantitative instruments, procedures	_____
6	Methods account for linguistic framing and differences in conceptual representation	_____
7	Methods used are comfortable to and effective for target users	_____
8	Methods can be implemented within the situational context of the target users	_____
9	Methods are implemented by culturally competent researchers or indigenous research assistants	_____
10	Methods are pilot-tested on a subset of target users	_____
11	Methods are as portable or as mobile as possible	_____
12	Data to be analyzed qualitatively can be transcribed into equivalent forms	_____
13	Data to be analyzed quantitatively is checked for distributional equivalence	_____
14	Data to be analyzed quantitatively is checked for reliability equivalence	_____
15	Data to be analyzed is disaggregated by cultural groups	_____
16	Results are translated using relevant cultural frames of reference	_____
17	Results are transformed to reflect the actual data patterns	_____
18	Results are generalized consistent with statistical or hermeneutic frameworks	_____
19	Results are verified by subject matter experts and indigenous representatives	_____
20	Results are shared with the target communities	_____

Source: Smith-Jackson and Essuman-Johnson, 2010; copyright permission pending.

need to increase the quality of medical research to support equity in the positive benefits of research, the US National Institutes of Health Revitalization Act (1993) mandated inclusion of underrepresented minority participants in clinical trials. This act also applied to the training of journal manuscript reviewers who were supposed to be educated to place value on the diversity of samples when reviewing and critiquing studies (Yancey, Ortega, and Kumanyika, 2006). The goal was to increase the applicability of medical studies to diverse populations, with the caveat that representative samples (per the population) may not be enough to allow valid comparisons. In most instances, oversampling of certain groups should be implemented.

However, regulations or policies to ensure inclusion do not necessarily guarantee inclusion will be practiced. Durant, Davis, St. George, Williams, Blumenthal, and Corbie-Smith (2007) conducted an analysis of data collected from 440 principal investigators (PIs) with grants to determine how well the studies met targeted enroll-ment and recruitment goals. Ninety-two percent of the PIs set goals to recruit African Americans, but less than one-third of the PIs met their goals. Sixty-eight percent of the PIs set goals to recruit Hispanics; 16% met those goals. It was similarly disheart-ening for participants of Asian descent; 55% of PIs set goals to recruit, but only 9% met goals. Native Hawaiians/Pacific Islanders (35% set goals; 0% met goals), and American Indians (23% set goals; 0% met goals) were also drastically underrepre-sented in spite of the goals of the PIs and the approval (and funding) of the sponsor-ing agency. One might argue that the absence of adequate representation should not have any significant negative consequences for generalizability of the results popula-tions not represented in the studies. These are naïve arguments and may also reflect a need to defend the status quo.

A concrete rationale can be given by focusing on pharmacology, a vast global industry sector that has achieved financial success on an immense scale—but at whose expense? Many ethno-pharmacology and ethno-psychopharmacology studies have demonstrated significant variability in dose responses between ethnic groups arising from genetic polymorphisms (Yasuda, Zhang, and Huang, 2008). These differences, if not understood and applied, may lead to underdosing, overdosing, or prescribing entirely ineffective medications with dire consequences. Although targeted enrollment is reported by researchers in NIH studies, there are still large numbers of clinical trial results reported with highly unbalanced and nondiverse samples. Similar problems exist in the United Kingdom. As an example, Mason, Hussain-Gambles, Atkin, and Leese (2003) discuss clinical trials and other medi-cal studies that have been published with minimal inclusion of underrepresented minorities. One shared factor in the underrepresentation of certain minorities in the United States and, for example, South Asians in the UK (the largest minority group in the UK), is in the difficulty of recruiting minority participants. One cause of these difficulties is the history of medical and scientific abuse and exploitation for the sake of advancing research in many affluent countries. Other factors include an inability of research teams (who may be predominantly majority cultures) to understand how to build trust and rapport with communities. Another factor is an inability to convey the important benefits and risks to certain groups. Finally, the lack of provisions to minimize barriers to participation related to transportation, work release, funding/compensation, and language also play a strong role in failures of our research enterprise to be inclusive. Certain factors related to in-group/out-group behaviors and perceived intentions between researchers and participants also play a role in recruitment. An easy solution is to ensure all researchers conducting actual data collection are either indigenous samples or subsamples in the study or, if not, all research team members are culturally competent. It should also be noted that, again, within education and training, researchers will not place value on over-coming these barriers and will not have a full understanding of the negative impacts on their research.

There are many other solutions to overcome the historical barriers to designing and implementing high-quality, inclusive research, and numerous training sessions, papers, and presentations have addressed methods to ensure adequate representa-tion of minorities. However, the drivers of the research enterprise must first end the practice of incentivizing researchers whose studies fail the test of inclusion, and as a consequence, have only limited rigor or quality.

METHODOLOGIES AND METHODS

The design of inclusive research methodology to produce results, requirements, or guidelines for inclusive products and systems is indeed a challenge. Researchers must first identify appropriate methodology, and then select the specific methods to use that fit within the methodology. The words *methodology* and *method* are not interchangeable, yet researchers use them interchangeably or in ways that are confusing to the reading audience. Methodologies are akin to philosophies of research and can be considered broadly as the investigative lens the researcher has chosen to view the problem (or if one is using action research, the lens chosen by the community or target group to address the problem). A methodology is usually a guideline that describes what will be done or the systematic steps toward problem scoping to data collection to analysis to interpretation, and finally but not always the case, research translation. Methods are the precise mechanisms (within the methodology) used to acquire data or indicators to study the problem. A stopwatch is an instrument used to support objective measurement of performance time. A questionnaire might combine ratings (quantitative) and open-ended questions (qualitative) to elicit opinions, beliefs, or other subjective responses from participants. Do not confuse subjective and objective methods with quantitative and qualitative methodologies, although each can be used to guide decision making about the organization of a research project (Figure 1.7).

A methodology can be divided into two broad categories: quantitative and qualitative. Both quantitative and qualitative methodologies make claims about knowledge and place value on certain methods. For example, quantitative methodologies typically embrace positivist and reductionist knowledge claims. This means the methodology rests on a belief that a phenomenon must be reduced to its smallest, measurable form before it can be systematically explored. As a result, the phenomenon must be explored in a controlled, contrived setting. Qualitative methodologies often make knowledge claims about achieving rigor and validity through holistic examination of the phenomenon as it occurs in the natural world without manipulation by the researcher and little or no control over the setting. These two examples are the extremes of the two categories; there is much that lies between the two. Most researchers are trained in one or the other, but new and progressive education and training is preparing researchers who are well prepared for global challenges by integrating both qualitative and quantitative methodologies, with combinations of methods to support the methodologies.

In the past two decades, researchers are better trained to understand and appreciate differences in epistemologies, methodologies, and methods. In particular, they have learned the value of mixing research activities to arrive at a more comprehensive understanding of complex phenomena. With the advent of mixed-methods approaches championed by Creswell (2003) and others, it has become more acceptable and valued to conduct research that utilizes multiple methods within one study, utilizing both quantitative and qualitative methodologies. The methods can be

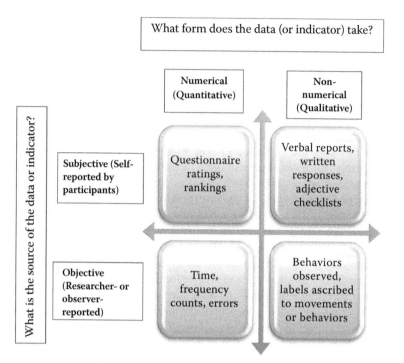

What form does the data (or indicator) take?

	Numerical (Quantitative)	Non-numerical (Qualitative)
Subjective (Self-reported by participants)	Questionnaire ratings, rankings	Verbal reports, written responses, adjective checklists
Objective (Researcher- or observer-reported)	Time, frequency counts, errors	Behaviors observed, labels ascribed to movements or behaviors

What is the source of the data or indicator?

FIGURE 1.7 Matrix with examples to distinguish quantitative and qualitative methodologies from subjective and objective methods.

arranged systematically to either sequentially or concurrently address the research problem (Creswell, 2003). For further discussions on these methods, including transformative research, see Creswell (2003).

Selection of the most appropriate research design does not guarantee the research will be inclusive. The methods themselves must be suitable for the target group. Numerous studies have been conducted that fail to generalize to anyone other than the homogeneous sample of participants who were used in the initial study and, more disturbingly, in the series of replications of the study that follow with yet another series of new samples of the same homogeneous participants. If, for example, a study was initially conducted with a sample of European Americans and their prevalence in the sample was 98%, any analyses that aggregate the sample values will most accurately reflect European Americans, the majority of the sample. Any outcomes will benefit European Americans, not others. To generalize results to a wider ethnic group from a design perspective is neither valid nor ethically responsible when a sample is overwhelmingly composed of one representative group. There may be exceptions to this rule, in that perhaps there are some metrics and indicators that are almost universal regardless of ethnic group membership. It is difficult to name any, because individuals are impacted both by genetic polymorphisms and their contexts or natural environments. Examples might be a simple target detection task or a finger-tapping task. One could safely assume that, barring any age-related or nutrition-related factors, simple target detection and finger tapping might generalize across multiple cultures. But studies such as these are numerous and have reached frequencies of replication that they are no longer necessary to implement.

The unrelenting influence of ecological factors can shape everything from anthropometric measures, such as reach envelopes, to preferences for icons, earcons, or hapticons on mobile devices. Most target groups for which we design will be diverse. The effort to ensure inclusion should outweigh any other considerations. The risk of designing products or systems that have unequal and negative impacts should override the convenience of conducting studies quickly and cheaply by utilizing only samples that are highly homogeneous.

In the quantitative arena, an additional problem is the use of central tendencies or aggregates to form the basis of descriptive and inferential statistics, and subsequently, researchers' interpretations of the results. When aggregated, these values become a point estimate that is usually closer to the most highly represented participants in the sample. If the participants' raw scores on, say, the dependent variable, do not converge in line with a cultural attribute, such as gender, ethnicity, class, educational level, or workplace experience, or generation, then the aggregation may not be a major issue in terms of culturally competent design (but could be an issue for other types of design). But, if the scores on the dependent variable are clustered and vary by cultural groupings of some kind, an aggregation may very well erase the important information that would have been revealed with closer scrutiny of the data. Take, for example, a hypothetical study that was conducted to understand usability of menus on satellite phones for military soldiers deployed to remote conflict zones. Using a sample of 10 users, 3 women and 7 men, a mean of 5.4 seconds ($SD = 2.22$ seconds) is calculated and represents the aggregated value of speed of locating the reconnaissance team channels. (The values are in seconds rather than milliseconds and are rounded to whole numbers.)

Let us also assume the researchers set a benchmark or criterion of 4 seconds as "success," and anything above 4 seconds requires a full redesign of the menu structure. With the results above, the designers will return to the drawing board because the finding of 5.4 seconds to locate a specific menu item is well above their success criterion. But, given the nature of the sample, there is an important piece of information that is overlooked because the researchers focused only on the aggregate. If we examine the data in Table 1.2, it becomes apparent there is some gender variation demonstrated by three of the participants who are women. The women participants performed faster than the seven men. The gender-based means were 2.67 ($SD = .58$) for the women and 6.57 ($SD = 1.40$) for the men. Certainly, the groups were unbalanced, which contributes to the problem, and could be corrected using other methods that "estimate" or weigh values. But most do not use these methods, because gender or other cultural factors are often ignored. The point to be made here is the degree of information lost in the aggregate, when cultural differences, in this case gender, are ignored. Perhaps the design was biased toward women, allowing for faster menu location through the use of certain types of verbal labels, menu levels, and structures, or spatial layouts on the display. Perhaps the menu design resembled certain devices that were more commonly used by women, and this familiarity allowed women participants to easily assimilate the new menu design with an existing mental model.

TABLE 1.2

Data from Hypothetical Study on Menu Design

Participant ID	Speed (sec)	Gender
1	3	F
2	6	M
3	2	F
4	7	M
5	4	M
6	6	M
7	3	F
8	8	M
9	7	M
10	8	M

However, by simply aggregating the data, deeper design analysis will not be possible. Likewise, without qualitative interviews, the differences that may be associated with gender will be overlooked and remain unknown. Most important, the cost of doing another study that may repeat the same oversights may be a loss to the company. A redesign of the menu might very well lead to an advantage only among users who are men, while reversing the advantage women had in the initial formative evaluation. One reason is that the designers moved away from whatever underlying attributes were providing advantages to women. Upon product release to the market, the final product may be biased, which might negatively impact usability, sales, and customer loyalty. There are a number of assumptions made in the use of this hypothetical case, but the goal is to suggest the need for careful scrutiny of the data.

Such practices as examining the distribution of values can lead to highly valid assessments of any data set and ensure the story that is ultimately reported by the researcher is as close to an approximation of the real world as possible. Additionally, researchers should be careful when using the increasingly advanced forms of data exclusion such as eliminating outliers. Outliers might very well represent a cultural group, and provide researchers with information about differential validity and reliability of their methods or instruments. Outlier elimination is really a lost opportunity to truly understand what is happening in the data. Researchers rarely examine the types or clusters of participants who are outliers, but simply inspect values for deviations from a set maximum or minimum in terms of the distribution of values. This common practice may consistently eliminate a set of participants who share common attributes found among the target users. It will also yield a "design for the average," which is usually a critical error in design practice.

Other problems arise from differences in the validity and reliability of instruments used in research. Similar to intelligence tests and other culturally biased instruments, the content, format, or manner of norming scores for specific instruments may yield differential validity or the differences may not be known because researchers norm the instruments on biased samples (e.g., predominantly male, predominantly White, upper socioeconomic status). For example, in a study of construction workers, one might measure the construct validity of a risk perception instrument by testing its correlation with another construct that should correlate with risk perception such as sensation-seeking behaviors. The two should be negatively correlated (higher risk perception is associated with lower sensation seeking), and the correlation should be strong. In most instances, researchers use all values of each construct in a correlation, and never conduct correlations by groups of interest to assess differential validity. It is possible that risk perception might have a strong negative correlation with sensation seeking, but only for non-Latino workers. The correlation using the data from Latino workers might yield only a low and nonsignificant correlation. Differential validity issues will impact further analyses, particularly when using regression, correlations, or analysis of variance. It is important to check for differential validity, since aggregating data hides the potential cultural differences, results in the loss of meaningful information, and produces unreliable statistical results.

Reliability of the instruments may not be the same across cultural groups. Reliability is the extent to which the same participant will report the same values over time, using the same or a similar instrument. Reliability is a measure of consistency, with the assumption that, within a certain margin of error, most participants will be consistent in their responses. This consistency can be tested using test–retest reliability (with the same instrument or an alternate form of the instrument), split-half reliability, or internal consistency reliability within the same instrument. Given the constraints on time for most studies, many researchers simply examine internal consistency, which reflects the degree of consistency a participant demonstrates on the full instrument (if unidimensional) or specific dimensions of the instrument (if multidimensional). Thus, if an instrument measures a construct using five items that are rated from 1 to 5, then we would expect a participant's ratings of the five items to be relatively consistent. We would also expect this same consistency across all participants in a sample. The reliability of scores on a questionnaire should always be checked before further analyses are conducted; otherwise, researchers have a GIGO problem (garbage in = garbage out). This is particularly important when working with heterogeneous samples. One related point is the need to always use a back-translation method that ensures the same mental representations are conveyed when an instrument is translated from one language to another language. Using at least two iterations of back-translation with individuals who are native speakers of the language is very important. Simply using direct translation may invalidate the results and will certainly lead to differential reliability—significant differences in reliability between cultural groups.

For an example of differential reliability, note the data from a study of Latino and non-Latino farmworkers conducted by Smith-Jackson, Wogalter, and Quintela (in preparation) in Table 1.3. In this study, we administered several questionnaires that had been piloted in a previous study and back-translated (Smith-Jackson, Wogalter, and Quintela, 2010). The "reliable" questionnaires were administered in a second study with a sample size of 109 participants (n = 54 Latino farmworkers; n = 55 non-Latino farmworkers) to explore the extent to which cultural differences in safety-related attitudes and perceptions influenced behavioral intent to avoid hazards;

TABLE 1.3

Differential Reliability Table

Construct	Total Sample	Latino (n = 52–54)	Non-Latino (n = 55)	Equivalent Reliability
Risk perception	.75 (6 items)	.54	.75	.79, .80 Two items retained
Safety self-efficacy	.81 (6 items)	.78	.80	All items retained
Supervisor's behavior	.77 (4 items)	.53	.71	.71, .79 Two items retained
Behavioral intent (dropped)	.28 (4 items)	X	X	X
Safety locus of control (dropped)	.47 (2 items)	X	X	X
Label (5 items)	.64	.72	.49	.57, .54 Two items retained
Brochure (5 items)	.13	.19	.24	.62, .50 Two items retained

Source: Smith-Jackson et al., in preparation.

ratings of usability of risk communications; and exposure to pesticides. The left-most column in the table is a list of the measured constructs in the study. The constructs of "label" and "brochure" reflected measures of the usability of two different risk communication designs. Cronbach's alpha internal consistency equation was used to measure reliability. Using Nunnally's convention (1973), a Cronbach's alpha value of ≥.70 is considered reliable and that level of reliability will also support statistical conclusion validity when conducting the subsequent analyses to support inferential logic for hypothesis testing. If we look at Risk Perception, the total sample reliability was r_{alpha} = .75 (with six items retained). But, when the sample internal consistency was computed by cultural group (Latino versus non-Latino), the reliability values of the Latino subsample is r_{alpha} = .54 and the non-Latino subsample was r_{alpha} = .75.

Upon further analyses by cultural group, it is clear that data from the Latino subsample produced low-reliability values (.54), which may be the result of several factors associated with internal validity. For example, although the back-translation verifications of mental representations were conducted, it is possible that the psychometric scaling attributes of the questionnaire (Likert-type ratings) was not fully matched to the needs and preferences of Latino farmworkers. Placing magnitudes on concepts or ideas is not a common practice of many cultures. Additionally, even though the items were read aloud in Spanish or English upon request, it is possible that random error in the way the items were read or delivered could have negatively impacted the responses of Latino farmworkers but had less of an impact on native English-speaking farmworkers. Likewise, other sources of systematic or random error could have undermined the internal consistency of the responses given by the subsample of Latino participants. With these results, it was important to go forward only with items that had equivalent reliability for both cultural groups. Further testing led to retaining only two items from the risk perception questionnaire with reliabilities of r_{alpha} = .79 for Latino workers and r_{alpha} = .80 for non-Latino workers. These items were retained and used in further analyses of performance measures such as knowledge transfer and time to locate information, as well as subjective measures of usability. Notice also in Table 1.3 that certain constructs were dropped altogether because initial reliability was low for both groups. In contrast, safety self-efficacy yielded high reliability equivalence for both groups (r_{alpha} = .79 and .80), so all items were retained.

If the reliability check was based solely on the item values for the full sample, the researchers would have assumed the reliability standard had been met, and all items would be included in the subsequent analyses (e.g., ANOVAs). Consequently, the true differences and patterns among both groups would have been masked by low-reliability items and the differential reliability between the groups would not have been discovered. Additionally, models developed on the basis of the items with low and/or differential reliability would fall into the GIGO category.

As shown in Table 1.4 (demographics), it is also important to understand the contribution of multiple group differences that might be layered within the analyses as contributing factors. For example, age and educational differences are obvious for both subsamples. These demographic factors should be analyzed and used as predictors/independent variables. Note that, for this study, half of the non-Latino participants were exposed to the EPA-standard design (US Environmental Protection Agency) of a pesticide warning label, and the other half was exposed to what was referred to as a culturally competent design of a pesticide warning label. The same was done for Latino participants. For all metrics used in the ANOVAs, a folded-F test was done to test for variance equivalence on all dependent measures by cultural group. Variance equivalence was supported for both groups. In the end, we found a pattern of differences across both groups but also some interesting similarities. For example, both Latino and non-Latino farmworkers demonstrated higher comprehension of the culturally competent label design compared to the standard EPA design.

In this study, as is often the case, it was equally important to examine the qualitative data, especially the responses to the follow-up interview questions on risk communication design, comprehension, and knowledge transfer after exposure to the brochures and labels. Participants, regardless of cultural group, preferred very

TABLE 1.4
Demographic Variables Grouped by Ethnicity

Variable	Latino	Non-Latino
Age	33.76 (10.94)	36.82 (16.82)
Education	6.99 (3.22)	12.89 (3.29)
Years experience	10.38 (7.19)	15.85 (13.10)
Symptom severity	23.02 (3.02)	26.66 (3.33)

simple, low-density text. Colors and graphics were very important. One important difference is that non-Latino participants reported a sense of empowerment to negotiate with farm owners to acquire protective equipment to prevent exposure to pesticides, while Latino participants did not convey this belief. Instead, Latino participants reported the need to arm themselves with as much information as possible to aid in self-protection, assuming they would not be protected by farm owners or supervisors.

SUMMARY

In this chapter, the necessity of cultural ergonomics was discussed and the drivers of this necessity were identified. The globe has always been diverse, but the transfer and exchange of products and systems across the globe has increased on an order of magnitude that imposes new demands on how we conduct research. Additionally, the rate of population shifts from poverty to affluence for many countries seems to be increasing exponentially. Access to products and systems has increased. Additional imperatives such as the prevalence of informal work systems, increases in migration, and secular trends within populations make it all the more obvious that the paradigm must shift and research practices must be made more explicit.

Advances in the knowledge domain have helped to disseminate cultural ergonomics practices, but the level of adoption remains slow. Adoption of innovations is always a challenge, especially when the innovations exist in a context where the complexity and relative advantages are not clear (Rogers, 1983). As the issue of change and the use of cultural ergonomics approaches gains prominence, providing frameworks and methods to researchers is critical. In this chapter, we offered three useful and broad frameworks, as well as several indicators or considerations to support inclusive research and design. It is unlikely the educational system will change quickly, but there are several institutions attempting to train and educate future researchers such that their practices are more inclusive.

REFERENCES

Akin, Ö. (1990). Necessary conditions for design expertise and creativity. *Design Studies, 11,* 2, 107–113.
Alostath, J., Almoumen, S., and Alostath, A. (2009). Identifying and measuring cultural differences in cross-cultural user-interface design. *Internationalization, Design, and Global Development. Lecture Notes in Computer Science*, Volume S623. Berlin, FRG: Springer.
Bandura, A. (1977). Self-efficacy: Toward a unifying theory of behavioral change. *Psychological Review,* 84, 191–215.
Bandura, A. (1982). Self-efficacy: Mechanism in human agency. *American Psychologist,* 37, 122–147.
Barber, W., and Badre, A. (1998). Culturability: The merging of culture and usability. Presented at the Conference on Human Factors and the Web, Basking Ridge, New Jersey: AT&T Labs. Retrieved from http://zing.ncsl.nist.gov/hfweb/att4/proceedings/barber/. January 6, 2013.
Barber, W., and Badre, A. (1998). Culturability: The merging of culture and usability. Graphics, Visualization and Usability Center (Georgia Tech). Proceedings of the At&T Labs. Online: http://research.microsolft.com/en-us/um/people/maryc2/hfweb98/barber/. Retrieved January 8, 2012.

Barab, S., Dodge, T., Thomas, M., Jackson, C., Tuzun, H. (2007). Our designs and the social agendas they carry. *The Journal of the Learning Sciences,* 16, 263–305.

Chapanis, A. (1974). National and cultural variables in ergonomics. *Ergonomics,* 17, 153–175.

Creswell, J. W. (2003). *Research Design: Qualitative,Quantitative, and Mixed Methods Approaches,* 2nd ed. Thousand Oaks, CA: Sage.

Durant, R., Davis, R., St. George, D., Williams, I., Blumenthal, C., and Corbie-Smith, G. (2007). Participation in research studies: Factors associated with failing to meet minority recruitment goals. *Annals of Epidemiology,* 17, 634–642.

Eckersley, M. (1988). The form of design process: A protocol analysis study. *Design Studies,* 9, 2, 86–94.

Ferreira, R. (2002). *Culture and E-Commerce: Culture-Based Preferences for Interface Information Design.* MS Thesis. Virginia Tech.

Frempong, G. (2011). Developing information society in Ghana: How far? *The Electronic Journal on Information Systems in Developing Countries,* 47, 1–20. Online: http://www.ejisdc.org/ojs2/index.php/ejisdc/article/viewFile/820/369

Greenwald, A. G., McGhee, D. E., and Schwartz, J. K. L. (1998). Measuring individual differences in implicit cognition: The implicit association test. *Journal of Personality and Social Psychology,* 74, 1464–1480.

Hall, E. (1966). *The Hidden Dimension.* Garden City, NY, Doubleday.

Hall, E., *The Dance of Life: The Other Dimension of Time.* Garden City, NY, Doubleday.

Hardy, S. (1993). The "Racial" Economy of Science: Toward a Democratic Future. Bloomington, IN, Indiana University Press.

Harel, D., and Prabhu, G. (1999). Global User Experience (GLUE). Design for cultural diversity: Japan, China and India, designing for global market. *Proceedings of the First International Workshop on Internationalization of Products and Systems,* pp. 205–216.

Hofstede, G. (1997). *Cultures and Organizations; Software of the Mind,* 2nd ed. London, UK: McGraw-Hill.

Honold, P. (2000). Culture and context: An empirical study for the development of a framework for the elicitation of cultural influence in product usage. *International Journal of Human-Computer Interaction,* 12(3&4), 327–345.

ISO/TR 7250-2:2010. *Basic Human Body Measurements for Technological Design—Part 2: Statistical Summaries of Body Measurements from Individual ISO Populations.*

Jhangiani, I. (2006). A cross-cultural comparison of cell phone interface design preferences from the perspective of nationality and disability. MS thesis. Virginia Tech.

Jhangiani, I. and Smith-Jackson, T. (2007). Comparison of mobile phone user interface design preferences: Perspectives from nationality and disability culture. *Mobility '07 Proceedings of the 4th International Conference on Mobile Technology, Applications, and Systems and the 1st International Symposium on Computer Human Interaction in Mobile Technology,* pp. 512–519. New York: Association for Computing Machinery.

Johnson, K. and Lichter, D. (2010). Growing diversity among America's youth: Spatial and temporal dimensions. *Population and Development Review,* 36 151–176.

Jones, J. M. (1979). Conceptual and strategic issues in the relationship of black psychology to American social science. In A. W. Boykin, A. J. Franklin, and J. F. Yates (Eds.), *Research Directions in Black Psychology,* pp. 390–432. New York: Russell Sage Foundation.

Jütting, J., Parlevliet, J., and Xenogiani, J. (2008). *Informal Employment Reloaded.* OECD Development Centre Working Paper 266. Online: http://www.oecd.org/dev/39900874.pdf. Retrieved December 3, 2011.

Klein, H. A. (2004). Cognition in natural settings: The cultural lens model. In M. Kaplan (Editor). *Cultural Ergonomics: Advances in Human Performance and Cognitive Engineering Research,* 4, 249–280. Elsevier Press Ltd.

Kroemer, K. H. E. (2006). *"Extra-Ordinary" Ergonomics. HFES Issues in Human Factors and Ergonomics Series,* 4. Santa Monica, CA: Taylor & Francis.

Lakoff, G. (1987). *Women, Fire, and Dangerous Things: What Categories Reveal about the Mind*. Chicago and London: University of Chicago Press.

Marcus, A. (February 2006). Culture: Wanted? Alive or dead? *Journal of Usability Studies, 1*, 62–63.

Mason, S., Hussain-Gambles, M., Leese, B., Atkin, K., and Brown, J. (2003). Representation of south Asian people in randomized clinical trials: Analysis of trials data. *British Medical Journal, 326*, 1244–1245.

Oh, C., Iridiastadi, H., and Smith-Jackson, T. (2011). Cultural critical incidents in design: An international perspective. In M. Gobel, C. Christie, S. Zschernack, A. Todd, and M. Mattison (Eds), *Human Factors in Organisational Design and Management–X, Volume 1*, Grahamstown, Republic of South Africa: Rhodes University, pp. 381–386.

Ossewaarde, M. (2007). Cosmopolitanism and the society of strangers. *Current Sociology, 55*, 367–388.

Paine, A. (undated). *Facing Issues of Quality in Collaborative Volunteering Research*. Session Four: Collaborative Practice in Inclusive Research Symposium. Online: http://ws1.roe-hampton.ac.uk/researchcentres/csvca/events/Angela%20Ellis%20Paine.pdf   Retrieved 8/1/2012.

Peters, G., and Peters, B. (February 2011). *Overdose!* Plaintiffmagazine.com.

Population Reference Bureau (2011). *The World at 7 Billion*. Online: http://www.prb.org/Publications/Datasheets/2011/world-population-data-sheet/data-sheet.aspx.   Retrieved August 9, 2012.

Rogers, E. (1983). Diffusion of Innovations. New York: Free Press.

Scott, P. (2009). Ergonomics in Developing Regions: Needs and Applications, Boca Raton, FL, CRC Press.

Shore, B. (1996). *Culture in Mind: Cognition, Culture, and the Problem of Meaning*. New York and Oxford: Oxford University Press.

Smith-Jackson, T., Iridiastadi, H., and Oh, C. (2011). Cultural ergonomics issues in consumer product design. In M. Soares (Ed.). In *Handbook of Human Factors and Ergonomics in Consumer Product Design: Methods and Techniques*. London, UK: Taylor & Francis.

Smith-Jackson, T., Brunette, M., Artis, S., Johnson, K., Perez, G., and Resnick, M. (2008). Self-globalization: Strategies in HFES education, research, and practice. *Proceedings of the Human Factors and Ergonomics Society 52nd Annual Meeting*, 662–666.

Smith-Jackson, T. L., and Essuman-Johnson, A. (2002). Cultural ergonomics in Ghana, West Africa: A descriptive survey of industry and trade workers' interpretations of safety symbols. *International Journal of Occupational Safety and Ergonomics, 8*, 37–50.

Smith-Jackson, T. and Essuman-Johnson, A. (2011). Culturally-competent ergonomics: A preliminary checklist. *Human Factors in Organizational Design and Management X (ODAM 2011, Vol. 1)*, Grahamstown, South Africa, pp. 387–392.

Smith-Jackson, T. L., Leonard, S. D., and Essuman-Johnson (2003). Comparison of USA and Ghanaian interpretations of prohibition and "to-do" symbols. *International Society for Occupational Ergonomics and Safety*.

Smith-Jackson, T. L., Nussbaum, M. A., and Mooney, A. M. (2003). Accessible cell phone design: Development and application of a needs analysis framework. *Disability and Rehabilitation, 25*, 549–560.

Smith-Jackson, T., Brunette, M., Artis, S., Johnson, K., Perez, G., Resnick, M. (2008). Self-globalization: Strategies in HFES education, research, and practice. Proceedings of the Human Factors and Ergonomics Society 52nd Annual Meeting, 662–666.

Smith-Jackson, T., Wogalter, M., and Quintela, Y. (2010). Safety eliminate and pesticide risk communication disparities in crop production. Human Factors and Ergonomics in Manufacturing and Service Industries, 20, 511–525.

Spallazzo, D. (2012). Cultural heritage and mobile technologies: Towards a design framework. *18th International Conference on Communication, Networking, and Broadcasting. IEEE.* pp. 108–116.

United Nations Population Fund (2011). *UNFPA State of the World Population in 2011: People and Possibilities in a World of 7 Billion.* Online: http://iran.unfpa.org/images/photo/EN-SWOP2011-FINAL.pdf. Retrieved July 6, 2012.

United States. Census Bureau Population Profile of the United States. Online: http://www.census.gov/population/www/pop-profile/natproj.html (retrieved May 10, 2012).

Waddell, L. (2010). How do we learn? African-American elementary students learning reform mathematics in urban classrooms. *Journal of Urban Mathematics Education, 3,* 116–154.

Washington, E., and McLloyd, V. (1982). The external validity of research involving American minorities. *Human Development, 25,* 324–339.

Weber, E. U., and Hsee, C. K. (1999). Models and mosaics: Investigating cross-cultural differences in risk perception and risk preference. *Psychonomic Bulletin, 6,* 611–617.

Willis, M. (1989). Learning styles of African-American children: A review of the literature and interventions. *Journal of Black Psychology, 16,* 47–65.

Yancey, A., Ortega, A., and Kumanyika, S. (2006). Effective recruitment and retention of minority research participants. *Annual Review of Public Health, 27,* 1–29.

Yasuda, S., Zhang, L., Huang, S-M (2008). The role of ethnicity in variability in response to drugs: Focus on clinical pharmacology studies. *Clin. Pharmacol. Therapeutics, 84,* 417–423.

2 Global Issues

Marc L. Resnick and Sharnnia Artis

CONTENTS

INTRODUCTION

Recent decades have shown a clear trend toward increasing globalization in many domains. World trade in merchandise alone totaled more than $12 trillion in 2010. The United States imported almost $200 billion in goods and services in October 2010 and exported almost $160 billion. Most of these products and services can benefit from the application of sound ergonomics principles in their design, evaluation, and delivery.

Countries also interact in the military domain. The United Nations, NATO, and other international alliances coordinate military exercises. Misunderstandings in these interactions can lead to serious cost, damage, and loss of life. Nonmilitary interactions, such as joint space exploration, also can lead to misunderstandings that jeopardize the success of international efforts. With the retirement of the US Space Shuttle program in 2011, US astronauts will be catching rides to space on launch vehicles designed by and for astronauts from countries that differ from the United States in a variety of dimensions.

The complexity of global dynamics makes these significant challenges. The United States has significant interactions with a variety of countries, each with a unique population, culture, and national practices. Each interaction has different requirements in terms of understanding user requirements, practices, and behaviors.

It is also important to keep in mind the warning of Matsumoto and Juang (2008) that countries are not homogeneous; they are composed of a variety of subcultures, climatic regions, and languages/dialects. Satisfying this diverse set of users requires a detailed understanding of how differences translate into product, service, and interaction design.

DIMENSIONS

There are many dimensions that can vary from one country to another. This chapter will review differences along several of these dimensions, highlight some applications that illustrate how differences manifest in user requirements, and suggest some design solutions that would satisfy local populations in the applicable countries. Future challenges are addressed later in this volume in Chapter 14. Some of the dimensions of note include:

- Language
- Anthropometry
- Climate
- Laws and regulations
- Economics and wealth
- Cognition
- Social interactions

LANGUAGE

The most obvious difference in designing for international populations is the need to translate an original language into the languages prevalent in each location where the organization wants to deliver a product or service. There are several levels at which language can be accommodated (Matsumoto and Juang, 2008). At the simplest level, the text can be translated word for word. This is the simplest and least expensive solution but can lead to problems when connotations of specific terms differ from one country to another. Amusing examples abound for assembly instructions that are translated from Chinese into English. This approach also fails to accommodate the user interface problems that arise when the words in a language are significantly longer or shorter than the original. For example, when translating a dialog box from English to German, it is often required to lengthen all of the text input boxes and labels. At the least, this requires extending the size of the dialog boxes. In many cases, it requires fundamental changes in the design of the dialog because they no longer fit within the window boundaries.

The second level is to translate the syntax and grammar along with the lexicon. This takes a little more effort and has a somewhat better result. This is the level that current state-of-the-art automated translations systems have reached as of the publishing of this chapter. Even better is to translate the semantics of the language. In this case, the meanings of the words are considered, and alternative terminology is selected if it better matches the intention of the text. Entire paragraphs may be rewritten to accommodate syntax and semantics simultaneously. At this third level, users in a target country can understand the meaning of the translated text, although effort may be required to overcome deeper cultural differences.

The fourth level considers the pragmatics—a complete rewriting of all text to customize it for the conceptual, cognitive, and social differences from one location to another. This resolves cultural idiosyncrasies, colloquial interpretations, and differences in the connotation of ambiguous terms. It requires a fundamental understanding of both the source language and the target language at the cultural level. It is also considerably more costly and time consuming to implement.

Tiewtrakul and Fletcher (2010) provide an extensive set of examples of language translations that failed to meet the expectations of the designers. One of the primary areas where language translation has been studied is in international air travel. These challenges arise because the cockpit crew are native speakers of either the language at the country of origin or at the language of the destination. So in many flights, the cockpit crew will speak a different language than the air traffic controllers at either the origin or the destination. Sources of error include accents that distort understanding of communications, cultural terminology, sentence structure and syntax, and idiomatic expressions. Challenges can range from slower comprehension that delays responses to misunderstandings that can lead to crashes or runway incursions. They present solutions such as developing communication protocols that reduce the use of jargon and formalize a single syntax that is used by all crews regardless of their native language. They also suggest training crews to enunciate clearly. However, recalling these protocols and formalizations increases the mental workload of the crew and can increase risk in some cases.

ANTHROPOMETRY

Pheasant and Haslegrave (2006) include an extensive set of databases of anthropometric data for several international populations. Even a cursory review of these data demonstrates large differences in sizes between residents of different countries. Parsons (1995) discusses the importance of having data for international populations to effectively design work systems, including control room design. Imrhan (2010) compared hand anthropometry between a Bangladeshi population and other international databases and found differences as large between male Bangladeshi populations and other male populations as there are between Bangladeshi males and females. He speculates on the implications of these differences in hand tool design, machine access spaces, and handheld devices. Agha (2010) investigated the consequences of anthropometric differences when importing school furniture into the Gaza Strip and found significant mismatches.

In a more focused study, Resnick and Corredor (1995) illustrated the perils of ignoring anthropometric differences when designing or installing equipment. They investigated the consequences of using industrial equipment designed for the US population in a Colombian workplace. Using height as an example, they found that designing for 90% of the US population would accommodate only 58% of the Colombian male population and 68% of the Colombian female population. Using the US data for elbow height, the dimension commonly used to set work heights, to set the height at which workers do their work over an 8-hour day would result in extreme wrist postures for the Colombian workforce. Using the US data for shoulder height to set the upper level of shelving or other reaches would result in extreme shoulder postures for the Colombian workforce. They conclude that attention to national differences in anthropometry is critical for maintaining the safety of the workforce. Similar consequences could be extrapolated to consumers and other user categories when applying anthropometry to design.

CLIMATE

One might not immediately realize the implications of climate on the application of ergonomics. However, there are documented examples that illustrate its relevance and ergonomists ignore climate at their peril. For example in hot and humid climates, users of cell phones need to exert a greater grip force to hold onto their mobile phones to keep them from slipping out of their hands. Designers of cell phones in India have addressed this challenge by using materials with greater surface friction in the design of the casing. Alternatively, they can use more textured grip designs, rubber skirts, or other solutions that enhance grip.

Similar requirements emerge in cold climates. When users might be wearing gloves when manipulating controls, the design must be reconsidered to accommodate reduced motor coordination, tactile sensitivity, and other capabilities. Anyone who has tried to use their mobile phone wearing gloves has experienced this decrement. Compounding this problem, many touch screen technologies cannot be used with gloves because they rely on a signal that is blocked by the glove material. Several unique solutions have been generated to this challenge, including the use of a sausage casing in South Korea, which some enterprising innovator discovered can reproduce the signal to a more high-tech design that integrates a special material into the thumb and forefinger of gloves that enable it to interface with touch screens (Greer, 2010).

LAWS AND REGULATIONS

There are some fundamental differences in how countries create and apply laws and regulations. In the practice of many subdisciplines of ergonomics, these differences can be critical in determining how practitioners should apply their craft. In the British system, which is used by the United States, general principles in legislation are further fleshed out in the promulgation of regulations and even further in case law coming through the courts. The French system puts more responsibility on initial legislation to specify the requirements right at the start. Understanding these differences is important in the design of many product categories as well as the practice of human factors forensics.

Another difference lies in the degree to which laws and regulations are enforced. In some cases, lack of resources prevents countries from enforcing their laws. In other cases, corruption prevents effective enforcement. In both cases, there is a condition of *caveat emptor*. Ergonomics can help the situation by making products safer and easier, even in the absence of legal and regulatory requirements to do so.

A simple example is the ban on the use of the swastika in Germany for media products such as video games. This required US video game companies to remove these from their World War II first-person shooter games. To identify these requirements, companies need to monitor the laws and regulations in any country in which they plan to do business. Countries that have stricter rules for violence, sexual content, or adult themes also must be accommodated (Resnick, 2006).

A more complicated example of a difference in legal requirements that has implications for ergonomics is in privacy law differences between the United States and the European Union (Milberg, Smith, and Burke, 2000). Privacy laws are much stricter in the EU. This has led to legal problems for companies such as Google in the delivery of customized ads, Amazon in the presentation of customized product recommendations, and Facebook in the collection of personal information from user profiles and posts. These companies have been forced to create different information architectures and process designs for the EU versions of their web sites to accommodate the stricter privacy laws. Additional functionality is needed to allow users to set privacy profiles.

ECONOMICS AND WEALTH

To redesign products for countries with less wealth, it may seem that providing bare-bones designs with fewer functions and lower quality would be effective. However, this has been shown not to be the case. These countries need different functionality rather than simply less. For example, the uncertainty of electrical power availability in countries such as India led mobile device makers to include longer-lasting batteries rather than shorter. Some also added a flashlight feature that was not available in the United States. Higher illiteracy rates mandate the use of more icon-based interfaces that may require significant user interface design changes to navigation menus, dialogs, and product descriptions.

Differences in economics also can require fundamental differences in functionality. The lack of a banking infrastructure in rural India led to the development of a new business model to supply residents with banking-type services without having bank accounts or bank branches (Karugu and Mwendwa, 2007). This adjustment required a fundamentally different user interface for the mobile devices that were used to deliver these mobile banking services. The higher levels of corruption in some developing country markets also mandates changes in the business delivery process and associated changes to the user interface design, with more verification and validation functionality.

COGNITION

Research on the physiological basis of cognitive differences is mixed at best (Matsumoto and Juang, 2008), but there are many cognitive differences that emerge from differences in experience, education, and training. For example, internationally customized websites are more likely to be usable because the match between the design and the users' schema reduces the cognitive effort needed to process information, navigate, and otherwise conduct efficient interactions.

Matsumoto and Juang (2008) describe an amusing but illustrative example of different situational processing approaches between international populations. A study hypothesized that western and aboriginal populations would have different cognitive capabilities to apply logic. They presented participants with logical syllogisms such as "All children like candy. Mary is a child. Does Mary like candy?" The difference turned out not to be a difference in logic cognition but practicality. Responses of the aboriginal cohort included a statement such as "How should I know? I've never even met her!"

Sheikh, Fields, and Duncker (2010) present an experience-based example of direct relevance in ergonomics. They conducted a card sort with British and Pakistani participants using a corpus of grocery items. Only translating the words from one language to the other in the navigation menus, but maintaining the same information architecture, was ineffective. There were differences in the breadth/depth trade-off between the two sets of users, and the Pakistanis used a religious organization to accommodate halal laws. Navigation performance was significantly better when the organization of the grocery inventory was reconstructed to accommodate these differences.

A difference due to fundamentally different educational approaches was reported in Matsumoto and Juang (2008). Chinese and US children both memorized simple math using a rote learning technique. But for the Chinese children, the memory modality was spatial, whereas for the US children, the memory modality was semantic. They hypothesized that this could be due to Chinese students' use of an abacus for practice and US students' use of arithmetic tables. There is little research that investigates how this modality difference would impact use of spatial versus semantic problem solving, display processing, or other system interactions in applications such as cockpit design later in life.

Lee (2010) investigated the impact of nationality on the perception of warnings. Using a warning for a low overhand hazard, he found that Korean participants focused on the event (the potential impact of one's head with the overhang), whereas the US participants focused on the consequences (potential injury and pain). Subsequent user testing found higher ratings from participants who matched the designer's nationality than those that mismatched. He hypothesized that this was based on the cultural differences whereby people in the United States have a more egocentric view due to higher levels of individualism and focus on the consequences for themselves. The Koreans are more group oriented, and so they focus on the aspects of the hazard that are true for everyone. Similarly, Chan et al. (2009) found that US participants interpreted symbols designed by US designers better than Hong Kong participants did. As a result, they recommend integrating cultural investigation into the ANSI process for warning symbol development.

SOCIAL INTERACTION

The most widespread model that describes differences in social interaction is that of Hofstede (1980). He warns that this model is focused on culture rather than on nationality and that many countries have a diversity of cultures therein. But for simplicity, most of the research in this area has sampled from national populations or focused on the dominant culture within a country. Hofstede's model has four dimensions of culture:

- Individualist–collectivist: Individualist cultures put a higher priority on individual goals than on group goals. Collectivist cultures put a higher priority on group goals. The definition of group depends on the structure of the group and can range from one's immediate family, to the extended family to an entire ethnicity. It can also manifest through organizational cultures from team, to facility, to company-wide allegiance.

- Power distance: High-power-distance cultures believe that people merit power, respect, and authority by virtue of their position in an organizational or national hierarchy. Low-power-distance cultures believe everyone is equal regardless of position, and hierarchies are just for efficiency of operation.
- Masculine–feminine: Members of masculine cultures are more ambitious, assertive, achievement oriented, and materialistic. Members of feminine cultures are more social, group oriented, and nurturing.
- Uncertainty avoidance: Members of low-uncertainty-avoidance cultures are more likely to be risk takers and more easily accept ambiguity. Members of high-uncertainty-avoidance cultures avoid risk, prefer structure and security, and avoid competition and conflict.

These dimensions must be taken with a grain of salt because there is so much variation within any culture. While members of low-uncertainty-avoiding cultures tend to be more risk taking, there are certainly many nonrisk takers in the population. The same is true with each of the other dimensions. It is an error to overgeneralize and design only for the exemplars of each culture. However, these dimensions can be used to guide some basic design decisions as long as they are followed up with valid user-centered methodologies.

Zandpour et al. (1994) found that advertising that reflects a societies' values is more powerful, perhaps because it more closely matches the recipients social schema and is easier to process. A better match to user values may explain why the EU has chosen to regulate Internet privacy more strictly than the United States (Milburg, Smith, and Burke, 2000).

INDIVIDUALIST–COLLECTIVIST

One way that this dimension manifests is in the design of worker incentives (Resnick, 2007). Individualist workers who are common in the United States are motivated by individual incentives that create competition. In fact, team incentives are often needed to ensure that extreme competition doesn't become counterproductive. However, in collectivist cultures, even being perceived as competitive can be embarrassing and would be a disincentive to work harder. Team incentives not only work better but are essential. Furthermore, they should not be framed as a bonus for hard work compared to others. Instead, the threshold needs to be set objectively and work better when they are framed as due to luck rather than individual achievement.

POWER DISTANCE

Marcus and Gould (2000) discuss how power distance can be related to the graphical organization and linguistic aspects of a website. They found more structure, deeper hierarchy, more use of national symbols, more use of verification and validation services, and more access restrictions for sites oriented toward high-power-distance users. They also found greater focus on presenting the leadership structure associated with the site or the organization behind it (see also Robbins and Stylianou, 2008). High-power-distance cultures are more influenced by metamoderation—putting more faith in reviews and reviewers that are ranked highly within a reputation management system.

MASCULINE–FEMININE

There is very little research applying this dimension to ergonomics-related issues. One can hypothesize that masculine cultures are more likely to be attracted to pure play e-commerce sites, whereas feminine cultures may prefer the emerging socially oriented commerce models present on Facebook and Groupon. Members of masculine cultures may also be more capable on virtual teams where assertiveness is essential to be heard.

UNCERTAINTY AVOIDANCE

Members of high-uncertainty-avoidance cultures are more comfortable in structured environments with fewer choices, less ambiguity, and more security. This has clear implications for information architecture. Brugman, Kitchen, and Williams (2006) found that high-uncertainty-avoidance cultures prefer websites with fewer options and more structure, represented by a deep, narrow architecture. It is also more important to have clear, unambiguous labels and therefore designers should invest more resources on developing effective labels (also see Resnick and Sanchez, 2004). In contrast, members of low-uncertainty-avoidance cultures are more willing to browse through an architecture and try out unfamiliar features. They may be more willing to click on cross-selling recommendations or intercategory links.

DETAILED EXAMPLE: DESIGNING SOCIAL NETWORKS

The choice of whether and how to use social networking systems is heavily influenced by culture (van Vliet, Huibregtse, and van Hemen, 2010). For example, members of cultures high in uncertainty avoidance are less likely to use these systems in general because they are less trusting of the identity of other members' profiles and of the information contained in their posts, recommendations, and direct communications. They are also less likely to post their own content for fear of possible consequences such as identity theft. When they do share, they prefer tighter networks composed of only real contacts and with strict privacy controls and authentication.

Members of low-power-distance cultures, on the other hand, may be more likely to use social networking systems and to post content. Because they see others as equals, they are more open to sharing. Similarly, they are more influenced by the recommendations of other users because they are more likely to consider these opinions relevant to their own requirements. Members of individualistic cultures have greater adoption of social networks, perhaps they are less tied to their real contacts.

However, there is little research on how these dimensions interact. Would a member of a high-uncertainty-avoidance, individualistic, low-power-distance culture reflect the values of the uncertainty avoidance dimension, the individualist-collectivist dimensions or the power distance dimension? Or do they interact in complex ways that are unpredictable without detailed research?

The size and shape of the networks that users create also depends on these social interaction dimensions of culture. Members of high-uncertainty-avoidance and collectivist cultures will have tighter networks composed of a greater proportion of real contacts. But within these networks, they may post more content than their low-uncertainty, individualist counterparts.

Wan, Kumar, and Bukhari, (2008) suggest that invitation-only social networking systems are more effective at satisfying collectivist, high-uncertainty-avoidance cultures. This creates a safer environment at the expense of fewer external social or professional networking opportunities. These closed networks are largely used to facilitate real-world relationships rather than to create new ones.

The functions that social networking system users employ also are influenced by social interaction dimensions of culture. Marcus and Krishnamurthi (2009) found that members of cultures high on power distance make more use of authentication checks and privacy controls. Perhaps counterintuitively, Chapman and Lahav (2008) found that Chinese users, a typically collectivist culture, were more likely to discuss personally sensitive issues with strangers than French users, a typically individualist culture. They speculated that because collectivist users are unwilling to reveal embarrassing information within their strong real networks, social networking systems may provide an outlet that is unavailable otherwise. Takahashi (2010) discovered that Japanese youths joined multiple social networks so that they could differentiate their interaction strategies accordingly. They assumed different personas, depending on their intentions.

Finally, social interaction dimensions influence users' preferences for how social networking interfaces are designed. Members of high-uncertainty-avoidance cultures prefer fewer choices and more process tunnels and wizards rather than an expanded information architecture (Marcus and Krishnamurthi, 2009). Members of collectivist cultures do not want their profile pictures to be visible to nonmembers of their tighter networks. In fact, Marcus and Krishnamurthi (2009) report that collectivist users are more likely to use profile pictures that are not personal photographs to avoid drawing attention to themselves. Members of masculine cultures prefer more options in settings, navigation, and features.

Other dimensions of international customization have been less extensively studied with respect to social network design. Extreme climates, whether hot or cold, may increase social network use in general because of the inconvenience of interacting in person. In this case, facilitating interactions among real-world contacts should be the focus of design. For remote locations, the challenge for users and the focus of design may be to establish connections with new contacts.

SUMMARY

The purpose of this chapter is to introduce some of the issues related to culture and international customization. Seven dimensions that have been demonstrated to drive the need for this customization are highlighted and the related research is summarized. The implications should be kept in mind in reading the remaining chapters in this volume.

REFERENCES

Agha, S. R. (2010). School furniture match to students' anthropometry in the Gaza Strip. *Ergonomics*, 53, 5, 344–354.

Burgmann, I., Kitchen, P. J., and Williams, R. (2006). Does culture matter on the web? *Marketing Intelligence and Planning*, 24, 1, 62–76.

Chan, A., Han, S., Ng, A., and Park, W. (2009). Hong Kong Chinese and Korean comprehension of American security safety symbols. *International Journal of Industrial Ergonomics*, 39, 5, 835–850.

Chapman, C. N. and Lahav, M. (2008). International ethnographic observation of social networking sites. *Proceedings of CHI 2008 Association of Computing Machinery*: New York. 3123–3128.

Greer, C. (2010, December 10). Texting your gloves off this winter? Try Smart Touch gloves. RCR Unplugged blog. Retrieved on 1/11/11 at http://unplugged.rcrwireless.com/index .php/20101210/lifestyle/5885/texting-your-gloves-off-this-winter-try-smart-touch-gloves.

Hofstede, G. (1980). *Culture's Consequences: International Differences in Work-Related Values*. Beverly Hills, CA: Sage Publications.

Imrhan, S. N., Sarder, M. D., and Mandahawi, N. (2010). Hand anthropometry in Bangladeshis living in America and comparisons to other populations. *Ergonomics*, 52, 8, 987–998.

Karugu, W. N., and Mwendwa, T. (2007). Case study: Vodafone and Safaricom: Extending financial services to the poor in rural Kenya. United Nations Development Programme (UNDP). Retrieved on 12/10/10 at http://*growinginclusivemarkets.org*.

Lee, Y. S. (2010). Identifying similarities and differences of pictorial symbol design and evaluation of two culturally different groups. In *Advances in Cross-Cultural Decision Making*. D. Schmorrow and D. Nicholson (eds). Boca Raton, FL: CRC Press.

Marcus, A., and Gould, E. W. (2000). Cultural dimensions and global Web user-interface design: What? So what? Now what? In *Proceedings of the 6th Conference on Human Factors and the Web*. Retrieved 11/2/10 at www.amanda.com/resources/hfweb2000/ hfweb00.marcus.html

Marcus, A., and Krishnamurthi, N. (2009). Cross-cultural analysis of social networking services in Japan, Korea, and the USA. *Proceedings of the 3rd International Conference on Internationalization, Design and Global Development*. Human Computer Interaction International. 1–16.

Matsumoto, D., and Juang, L. (2008). *Culture and Psychology*. 4th edition. Wadsworth Belmont, CA: CENGAGE Learning.

Milberg, S. J., Smith, H. J., and Burke, S. J. (2000). Information privacy: Corporate management and national regulation. *Organization Science*, 11, 1, 35–57.

Parsons, K. C. (1995). Ergonomics and international standards. *Applied Ergonomics*, 26, 4, 239–247.

Pheasant, S., and Haslegrave, C. M. (2006). *Bodyspace: Anthropometry, Ergonomics, and the Design of Work*. Boca Raton, FL: CRC Press.

Resnick, M. L., and Corredor, O. (1995). Estimating the anthropometry of international populations using the scaling estimation method. *Proceedings of the 39th Annual Meeting of the Human Factors Society*. Human Factors and Ergonomics Society: Santa Monica, CA.

Resnick, M. L., and Sanchez, J. (2004). The effects of organizational scheme and label quality on task performance in product-centered and user-centered retail web sites. *Human Factors,* 46, 1, 104–125.

Resnick, M. L. (2007, February). Incentives—what rewards work best? *Industrial Safety and Hygiene News*, pp. 1, 20.

Resnick, M. L. (2006). Risk communication for non-physical hazards. In M. Wogalter (ed.) *The Handbook of Warnings*. Mahweh, NJ: Lawrence Erlbaum Associates.

Robbins, S. S., and Stylianou, A. C. (2008). A longitudinal study of cultural differences in global corporate web sites. *Journal of International Business and Cultural Studies*, 3, 1–16.

Sheikh, J. A., Fields, B., and Duncker, E. (2010). Multi-culture interaction design. D. Schmorrow and D. Nicholson (eds), In *Advances in Cross-Cultural Decision Making*. Boca Raton, FL: CRC Press.

Takahashi, T. (2010). MySpace or Mixi? Japanese engagement with SNS (social networking sites) in the global age. *New Media and Society*, 12, 3, 1–23.

Tiewtrakul, T., and Fletcher, S. R. (2010). The challenge of regional accents for aviation English language proficiency standards: A study of difficulties in understanding air traffic control-pilot communications. *Ergonomics*, 53, 2, 229–239.

van Vliet, T., Huibregtse, E., and van Hemert, D. (2010). Generic message propagation simulator: the role of cultural, geographic, and demographic factors. In *Advances in Cross-Cultural Decision Making*. D. Schmorrow and D. Nicholson (eds). Boca Raton, FL: CRC Press.

Wan, Y., Kumar, V., and Bukhari, A. (2008). Will the overseas expansion of facebook succeed? *IEEE Internet Computing*, 12, 3, 30–34.

Zandpour, F., Campos, V., Catalano, J., Chang, C., Cho, Y. D., Hoobyar, R., Jiang, S., Lin, M., Madrid, S., Scheideler, H., and Osborn, S.T. (1994). Global reach and local touch: Achieving cultural fitness in TV advertising. *Journal of Advertising Research*, 34, 5, 35–63.

3 Role of Culture in the Design and Evaluation of Consumer Products

Shreya Kothaneth

CONTENTS

INTRODUCTION

This era of globalization has created a transnational exchange of people, trade, technologies, and innovations (Shome and Hegde, 2002). Consumers from different cultures have been reported to display different opinions, inclinations, and values and still continue to be hesitant to try new, foreign products (Suh and Kwon, 2002). This has serious implications in the consumer product domain where success is attributed by high sales of a product, and it is crucial that customers like what they see. For instance, Dell, which is an extremely successful computer manufacturer worldwide, has been found to have established only 5% of Korea's market share, and this pattern was reflected in China as well. Incidentally, Hofstede's (1984; 2003) research on national culture suggests that countries differ from one another based on four dimensions, power distance, individualism/collectivism, masculinity/femininity, uncertainty avoidance and long-term orientation.

Individualism refers to a preference for a flexible social framework, where people need to look after themselves whereas collectivism refers to an inclination where people expect their peers to take care of each other and their personal goals are often made keeping the entire group in mind (Hofstede, 1984). Hofstede's research (1984) also indicates that both China and Korea are considered to be highly collectivist societies. This difference has been hypothesized to be the reason behind the lack of interest in Dell's promoted feature on the ability to personalize the computer. Personalized computers end up being very different from the standard, and it was found that only individuals from individualistic countries prefer customizable products more than those from collectivist countries (Moon, Chadee, and Tikoo, 2008). Additionally, Chinese consumers have been known to emphasize products that facilitate group harmony (Neelankavil, Mathur, and Zhang 2000), while Europe and North America prefer products that help them express their individual personalities (Moon, Chadee, and Tikoo, 2008). A study by Nisbett (2003) on Easterners (defined as people from China, Korea, and Japan) and Westerners (defined as people from Europe, the United States, and the British Commonwealth) found that Easterners anticipate change and are less perturbed by it, whereas Westerners view their surroundings logically and are more surprised with an unexpected change. The study also found that when forced to deal with two contradictions, Easterners would seek to find some element of truth in both, whereas Westerners would reject one in favor of the other (Nisbett, 2003). Another study done to understand what usability meant to users from different cultures found that Americans related "usability" to the ability to use minimal resources and effort, while Germans related "usability" to ease of navigation. The Australian sample found ease of navigation to be important while the British sample felt the ability to achieve goals and navigation was important.

Consumption of consumer products are especially influenced by cultural attributes due to the difference in value placed on the products. What may be considered valuable to one culture may be extremely regular to another. For instance, cultural groups who emphasize empathy and security, will be most hesitant to readily adopt new products, while those who place value on hedonism will be very likely to adopt products that support their lifestyle (Daghfous, Petrof, and Pons, 1999).

Consumer products are also important beyond their actual use and value, due to their ability to communicate cultural meaning (Douglas and Isherwood, 1978). Cultural meaning has been has been considered to be found in three places; the world of culture, the consumer product, and the consumer (McCraken, 1986). Design has been reported to be most important factor when it comes to success of a product. Culture was found to directly impact efficiency of a user using a certain product (Wallace and Yu, 2009).

In 2007, Nokia grew to establish a 40% global market share and they shipped an impressive 437.1 million phones by the end of 2007. Nokia's growth was almost uniform in all the regions, barring one country—the United States. The total sales in

2007 went down by 23.3% in the United States (Epstein, 2008). By the end of 2009, Nokia's market share in the United States was only 7%. In contrast, more than a 100 million mobile phones were sold in India in 2009, out of which Nokia still established 54.1% of the market share (Krish, 2010). In fact, India overtook the United States and is now the second-largest market for Nokia. Nokia executives believe that India will continue to be high-end market since 81% of the market comes from high-demand urban areas (Tripathy, 2009). Nokia executives attribute the decline to the failure of customizing the phones to American end-user preferences, but instead producing devices for a global market (O'Brien, 2009). Other analysts concur with this statement attributed the loss of foothold in the United States to Nokia's one-size-fits-all policy, which ignores American customer needs and preferences (Hempel, 2009).

The infamous consequence of introducing snowmobiles to a small group of people in Finland, serves as another reminder as to why it is important to understand cultural attributes before launching technologies. Reindeer herding was the main occupation of the Skolt Lapp society, and the status of men was often judged by how well he/she herded deer. These people took pride in their reindeer and treated them extremely well. The snowmobiles provided a new mechanism for herding deer, but the warm relationship that once existed between the reindeer and herdsmen was ruined by the loud noises of the snowmobiles. The reindeer were terrified of these machines and ran away the minute they were aware of them being nearby. This eventually led to reduced number of calves being born each year. This resulted in mass unemployment and debt among the Skolt Lapps. Analysts who examined the impact of the snowmobile revolution, attributed the rapid diffusion of this unnecessary product that eventually ruined them, to them being an individualist group, and once the affluent adopted the snowmobiles, it rapidly became a common household item (Rogers, 1995).

BACKGROUND ON CULTURE

Culture is a complex term that has many dimensions (Evers and Day, 1997). It has been used to describe ethnic or religious societies, but can be used for other groups, such as families, organizations, etc. (Hofstede, 1984). According to a study on more than 50 countries, national culture can be measured in five dimensions: power distance, uncertainty avoidance, individualism/collectivism, masculinity/femininity, and long-term orientation (Hofstede, 1982, 1984, 2003). Power distance deals with the perception of inequality in society. It is defined as the measure of interpersonal power. Less-educated countries with lower-socioeconomic-status occupations were found to produce high-power-distance values (Hofstede, 1984). Countries like India and Mexico have a high-power distance, no matter what the occupation is, whereas Japan has a medium-power distance, across all occupations. Uncertainty avoidance is defined as the amount by which uncertain or new situations are perceived as a threat. High-uncertainty-avoidance cultures view new situations as dangerous and display a lower risk tolerance and often decline new ideas. Low-uncertainty-avoidance cultures like to consider themselves as risk takers and often readily accept new ideas. Individualism refers to a penchant for a flexible social framework, where people need to look after themselves, whereas collectivism refers to an inclination where people can expect their peers to look out for one another and their personal goals are often made keeping the entire group in mind (Hofstede, 1984). Masculinity is the amount by which a culture emphasizes separation of roles based on gender. Cultures with high masculine roles are usually male-dominated societies and prefer a more traditional lifestyle. High masculinity can be linked to competitiveness, aggressiveness and are often linked to more money-oriented desires, whereas high femininity can be linked to kindness and higher emphasis on home and children (Hofstede, 1997). Long-term orientation is the amount by which tradition or forward thinking is accepted or not. This dimension was added later to the four dimensions after it was discovered with an additional study on the West and the East. Cultures with higher long-term orientation scores tend to respect tradition and place value on long-term obligations. Further research led to the coinage of the terms *allocentrism* and *ideocentrism*, which are considered to be personality dimensions at the psychological level. They correspond to collectivism and individualism, respectively (Triandis, Bontempo, Villareal, Asai, and Lucca, 1988). Triandis (1994) also identified complexity, individualism, collectivism, and tightness as cultural syndromes (Triandis, 1994).

The cultural lens model also attempted to provide a framework to comprehend the notion and source of national culture (Klein, 2004). This model is based on the assumption that individuals who belong to the same national unit and who grow up in similar social environments tend to have similar experiences. It also believes that that these individuals have had similar childhoods that can be related to each other. These similar experiences, coupled with learning and impersonating, produce similar social, performance, and thought processes. The model attempts to capture these effects through dimensions that can offer a lens through which each individual observes the world around them. This lens acts as a filter and helps organize received information into communication patterns. When people have different backgrounds, they tend to think differently, and thus view the world differently. This opinion was shared by McCraken (1986) who believed that culture is the lens through which people viewed and understood different phenomena (McCraken, 1978, p. 72). Different perceptions of the world can cause trouble and disagreements during international dialogues (Klein, 2004). Other researchers have coined more dimensions to describe differences across cultures. For instance, a mastery proclivity is founded on the belief that people are in control of outcomes and that humans are dominant (Kluckhohn and Strodtbeck, 1961). They believe that anything is possible with adequate money, time, and cognition. Obstacles are viewed as signs to try other paths or methods. On the other hand, individuals who hold a fatalistic view believe that outside factors influence our lives. They endeavor to accept and adjust rather than attempting to solve the problem (Kluckhoh and Strodtbeck, 1961; Lane and DiStefano, 1992). *Time Horizon* is another such dimension that is used to describe when people set goals. Present-horizon indicates a preference for short-term goals whereas future-horizon indicates with worrying only about future goals (Adler 1991; Lane and DiStefano, 1992).

Another aspect of cultural research involves the role of gender. Gender has been recognized as a set of cultural roles (Lerner, 1986). As children become socialized, they start to link various personality traits to either being male or being female, adding it to their overall belief system. This association is recognized in consumer behavior research as gender identity because it is believed to have an impact on consumption behavior (Fischer and Arnold, 1994). For instance, masculine people (male or female) were found to be object focused whereas feminine people (male or female) were found to be people-focused (Palan, Areni, and Kiecker, 2001). Research has also shown that men refrain from shopping in general, because they perceive it to be a feminine activity (Danziger, 2005).

Age and/or generations can also have a huge impact on product acceptance. A comprehensive study on three generations of over three hundred families indicated that values were passed on from generation to generation (Hill, 1970). Results of this study also suggested that people from the same geographical location tended to have more similarities across generations. Older generations were also found to view and value luxury items differently than younger generations (Hauck and Stanforth, 2007). For instance, Generation Y (also referred to as 'Millenials') (Strauss and Howe, 1991) consumers enjoy spending more money on luxury items than older generations (Liang, 2005).

Socioeconomic status can also strongly influence consumption behavior. Consumers from lower socioeconomic statuses have been found to use luxury items more often because they are extremely aware of their status and are constantly trying to achieve a higher status level. However, the same study found that middle class consumers purchase luxury items not just to achieve a specific status but also to prove that they had the money to do so (Goldman, 1999).

Similarly, religion is another stratum that can influence consumption. Religions often have various guidelines like dietary restrictions, clothing, grooming, amongst others, that believers are expected to follow (Minkler and Cosgel, 2004). Small deviations from the guidelines can raise serious questions about commitment and religious belonging. For instance, a Muslim woman can indicate religious commitment with a scarf, but not with a beret or other accessory (Minkler and Cosgel, 2004). The advent of media in the twentieth century has led to more religious products. Religious art, cards, bumper stickers etc. are different ways that the sacred is professed (Hangen, 2001). Recent studies have found that religion impacts the use of communication technologies. Electronic bibles were found to be extremely popular amongst some mega church attendees. Voice over IP software was used to pray with people living in other countries (Wyche and Grinter, 2009). Calendars that cater to different religions are extremely popular and are now sold as commercial products (as shown in Figure 3.1). Similarly, orthodox Jewish families were found to purchase

FIGURE 3.1 Examples of different religious calendars that are commercialized and sold online.

automation technology to assist in Sabbath observation (Woodruff, Augustin, and Foucault, 2007).

BACKGROUND ON THE IMPACT OF CULTURE ON CONSUMER PRODUCTS

Consumer products encompass a wide range of products, from products used everyday like toothpaste to more expensive cars and electronics. For ease, consumer products are broadly classified into three main categories: convenience products, shopping products, and specialty products. This categorization has been used widely in the market domain (Lamb, Hair, and McDaniel, 2008). A brief summary of the differences between the types of products are shown in Table 3.1, and examples of cultural differences in consumption are described in the pages that follow.

CONVENIENCE PRODUCTS

Convenience products are considered to be the most popular products and are consumed and bought regularly. Convenience products are mainly purchased to save time and provide some comfort (Mooij, 2000). These products include everyday items like food, cleaning supplies, and personal care products (Lamb, Hair, and McDaniel, 2008).

A US survey found that 55% of the respondents said that convenience is "very important" when it came to purchasing food (Senauer, 2001). The increasing trend toward convenience also might be due to various paradigm shifts, such as working women, individualism, etc. (Scholderer and Grunert, 2005). Collectivist cultures have been found to invest more time and effort in preparing food, because they believe that consumption of meals has social meaning and value (de Mooij and Hofstede, 2002).

A study on women's fashion magazines in different parts of the world found clear differences between the products that were popularly advertised (Frith, Shaw, and Cheng 2005). They found that they could relate these differences to cultural differences in perceptions of beauty. For instance, porcelain skin is admired in China,

TABLE 3.1
Summary of Differences between Types of Consumer Products

Type	Convenience Products	Shopping Products	Specialty Products
What they are	Most popular; consumed on a regular basis; not very expensive	Less popular than convenience products; consumed frequently, but not regularly; medium price range	May be consumed as much as shopping products; considered to be luxury items; most expensive;
Examples	Food, personal care products, cleaning supplies	Clothing, electronics, household ornaments	Luxury automobiles, designer clothes

FIGURE 3.2   Examples of popular toothpastes sold in Asian countries.

whereas scarring the skin is a beautification process in Africa (Gengenbach, 2003). Popular beauty products advertised in Taiwan and Singapore were found to be those that were meant to enhance women's skin, hair, and face, whereas "clothing" was considered to be dominant in advertisements in the United States. This suggests that beauty in the United States is defined in terms of "the body," whereas beauty in the Asian countries is defined in terms of "the face" (Frith, Shaw, and Cheng, 2005). Creams and lotions meant to brighten the skin are still popularly sold in India, where fair skin is considered to be attractive. A study on toothpastes found that people from India, France, and Brazil chose toothpastes based on the least amount of artificial ingredients it contained (as show in Figure 3.2), whereas consumers in the United States found taste and ease of purchase to be the most important factors (Green, Cunningham, and Cunningham, 1974).

Japan has a high-uncertainty-avoidance index (Hofstede, 1983), and this is reflected in the high amount of antiseptics present in telephones, toothpaste, underwear, and bicycle handles (Becker, 1997). High-uncertainty-avoidance indices have also been reflected in the high consumption of mineral water, soap, and cleaning products (de Mooij, 2000). Asian countries have also been known for their use of natural products for beatifying and grooming purposes. For instance, coconut oil has been widely used for hair, skin, and overall health in India. It is so popular that it is even packaged and sold as a commercial product (see Figure 3.3).

SHOPPING PRODUCTS

These products are slightly less popular in terms of purchase frequency as compared to convenience products (Lamb, Hair, and McDaniel, 2008). Consumers spend more time researching and looking for these products since they are not only more expensive, but they also tend to include psychological value for the customer such as boosting their perceived self-image within a social group. Examples of these products include clothing, electronics, and household ornaments. Individualistic cultures were found to pay more attention to products for pleasure and excitement than collectivist cultures (de Mooij, 2000). A study on the use of German-based washing machines by Indian housewives found many interesting results (Honold, 2000). India has been considered to be a polychronous culture, which implies that multitasking is a common feature (Hall, 1959). This characteristic was attributed to the reason behind the criticism of the length of time a single wash cycle took (Honold, 2000). In India, clothes were frequently washed during the day, and it was rare to let unclean clothes sit for more than two days. Washing was also done in fixed time slots, and midmorning was seen as a popular time for most households. Washing was also done

FIGURE 3.3   Coconut oil packaged for commercial use in India.

in two loads at a time, separating the colors from the whites, so shorter wash cycles were necessary to get the cleaning done in their allotted time slots. Climate too was considered to be a huge impact on the type of preferred washing machine. Humid places required machines built of materials that do not rust. Another cultural difference was noticed by use of trolleys under the top-loading machines. The German engineers presumed that it was due to the smaller household structures, and the trolleys enabled the machines to be moved around from a bigger room, to a smaller area in the house with the necessary pipe connections. In reality, the trolleys were used to allow sweeping or cleaning under the washing machine. Windows were frequently kept open in houses in India, which also meant that the floors had to be swept every day to prevent accumulation of dust. There were also many product-design-specific features that were misunderstood by the Indian users. Fabric which required little or "easy care" was made out of synthetic material, which often didn't require ironing. These types of fabric are pretty popular in Germany so the machines were equipped with an "easy care" washing profile that could be activated by a button. Indian users either had no clue what the button was for or completely misunderstood it, using it wrongly for other types of fabric. There were also different wash cycle requirements due to the different properties of fabrics used in Indian clothing that were not included in the German washing machine. This implies that it is important for the background of the culture and environment must be understood before introducing a product.

Another product that has been designed keeping a specific culture and its preferences in mind are hot water boilers. These are extremely common in East Asian countries where tea drinking is a common feature and often a part of the meal. These boilers are now being increasingly sold abroad and are especially used by Asian American families (as shown in Figure 3.4). Similarly, specific rice cookers with features for different types of Asian-styled rice-based dishes, such as sushi and rice cakes, are also sold for use abroad by Asian families (as shown in Figure 3.4).

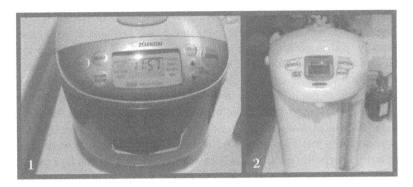

FIGURE 3.4   1. Rice cookers specific for Asian-styled cooking. 2. Automatic water boilers used in Asian homes.

When it comes to electronic products consumed worldwide, mobile phones espe-cially have diffused successfully, in spite of different cultures, values, and standards (Geser, 2004). The number of mobile phone subscriptions is said to reach five bil-lion by the end of 2010 ("Number of Cell Phones," 2010). Successful marketing of a product requires complete understanding of the markets, as well as methods of promotions (Iyer, Laplaca, and Sharma, 2006). Cross-cultural analysis and design characteristics need to be an integral part of the planning process, and manufac-turers will require design guidelines and checklists as globalization becomes more common (Marcus, 2002). Geographical conditions of the locations must also be kept in mind when designing phones. The phone needs to be resistant to heat, monsoons, humidity, dust, and bright sunshine. In countries like India, the tradition of eating with one's hands also makes it challenging for a designer to design a usable touch screen phone for use there (Lindholm et al., 2003). As already mentioned, Nokia has encountered problems with global acceptance of their mobile phones because they did not do the necessary background research (Hempel, 2009).

SPECIALTY PRODUCTS

These products are considered to be the most expensive product when compared to convenience and shopping products (Lamb, Hair, and McDaniel, 2008). Although these products may be purchased as often as shopping products, consumers are a lot pickier when it comes to them. Often, they are bought from retailers that provide the best deals. Examples of these products include luxury automobiles, high-end champagne, and celebrity-endorsed products. Having multiple watches and top-of-the-line cameras, as well as fashionable clothes, are also used as status symbols (de Mooij, 2000).

There are significant differences across cultures, when it comes to specialty products, because such types of luxury products are often used to represent success and power (de Mooij, 2000). According to Vatikiotis' study (1996), middle class consumers from Eastern Asia buy luxury items because they want to achieve a cer-tain level of status. Chinese consumers were found to differ from their American counterparts when it came to car purchases, as revealed in a study in 2000 (Johnson and Lieh-Ching, 2000). While American consumers were found to conduct extensive prepurchase research, Chinese consumers did not appear to follow suit. Additionally, with American consumers, income seemed to play an important role when it came to price negotiation. The higher the income, the less they negotiated. Whereas, Chinese consumers were found to negotiate irregardless of income. This could be related to the differences in the environment, and in particular car auto loans. Chinese consumers were found to prefer paying cash for their vehicles, since they didn't like the idea of the bank earning interest from them. Chinese consumers were also found to rely on opinions from friends and family a lot more than their American counterparts (Johnson and Lieh-Ching, 2000). This can be expected of a more collectivist culture such as China (Hofstede, 1983). Another study found that the social value of purchasing environmentally friendly or "green" products negatively impacted purchase intentions of hybrid cars by US consumers while it didn't have an effect on purchase intentions of Korean consumers (Oliver and Lee, 2010). Furthermore, cultures with a high masculinity index were found to purchase more luxury articles than those with a lower masculinity index (de Mooij, 2000). For instance, owning cheap watches was found to relate to lower incomes, but owning expensive watches wasn't found to be related to income, but instead was related to cultures with a high masculinity index.

THEORIES AND MODELS RELATED TO CULTURE AND CONSUMER PRODUCTS

There are two main types of cultural research in consumer behavior. One examines how cultural values of individuals influence a group's consumption behavior, and the other studies how the group's values impacts their individual's consumption behavior (David, Wong, and Tan, 1988). Sheth, Newman, and Gross (1991) developed the theory of consumption values. This theory suggests that consumer decisions are dependent on multiple values; that values have different impacts in different situations, and that consumption values are independent. According to this theory, there are five different types of values; functional value, social value, emotional value, epistemic value, and conditional value. Functional values are considered to be the most important value when it comes to consumer decisions. This value refers to the usefulness or utility of the product. Social values are considered to be influenced by the social image of a product. Emotional values are considered to stem from the ability of a product to evoke emotions. The romantic feeling that is stimulated by a candlelight dinner is an example of the emotional value. Epistemic values originate from the product's ability to evoke curiosity. Completely new and original experiences are considered to have high epistemic value. Trying a new brand can be considered to be a good example. Finally, conditional values depend on the situation and the environment. For instance, some products have seasonal value, such as Christmas trees, while others, like an ambulance, have value in emergency situations (Sheth, Newman, and Gross, 1991).

Various measurement scales have been developed to understand impact of cultural values. The Rokeach Value Scale, which is based on the Reokeach Value Survey (1973), is used to differentiate cultural groups as well as choice of products

(Musin and McIntyre, 1979). Similar scales include the Values and Lifestyles Scales (VALS) (Kahl, 1986) and the Chinese Value Survey, which was originally designed to complement the Rokeach Value Scale (Musin and McIntyre, 1979) for people living in locations where eastern cultural characteristics are prevalent (Bond, 1985).

Anthropological approaches have also been used to understand the relationship of culture and consumption behavior (David, Wong, and Tan, 1988). Advertisements and marketing gimmicks have been analyzed to understand different groups' values. For instance, a study on women's fashion magazines in different parts of the world found that beauty products advertised in Taiwan and Singapore focused on women's skin, hair, and face, whereas "clothing" was emphasized in advertisements in the United States (Frith, Shaw, and Cheng 2005). Similar approaches have been used by Belk and Pollay (1985), who conducted a longitudinal content analysis of popular US advertisements between 1900 and 1980 (Belk and Pollay, 1985).

Another model that is used popularly in consumer research is the theory of reasoned action by Ajzen and Fishbein, (1980) which is an extremely popular model that finds its origin in social psychology. It is based on the assumption that individuals usually behave in a rational manner, and they pay attention to the information around and consider the outcomes of their actions (Ajzen, 1985). According to this model, there are two main aspects that control behavioral intentions: a personal or "attitudinal" factor and a subjective "normative" factor. Subjective factors are the effects that the social surroundings have on their behavior. It is the individual's perception that people that matter are of the opinion that a behavior should or should not be performed. This can explain the relative unpopularity of cigarette consumption in India. India ranks only 118 in the list of smoking countries. However, chewing tobacco is still widely accepted and thus extremely popular ("Lighting up," 2009).

An extremely popular theory that has been used for consumer products is the semiotics theory. Semiotics is dedicated to the study and meaning of signs (Mick, 1986). Japanese packaging has been found to make use of *kanji*, a Chinese ideogram, to represent tradition and formality. *Hiragana*, which are simplified stroke characters, were found to represent femininity and tenderness, and *katakana*, which are meant to describe foreign words, are used to represent novelty and uniqueness of a product (Sherry and Carmargo, 1987). A semiotic perspective was also found to be helpful in understanding consumer behavior (Holman, 1976). Aesthetic signs and symbols comprise of colors, shapes, and materials, often used in marketing and advertising of consumer products. For instance, certain tribes in South Africa view red as a sociable color, while they perceive green to be hostile. This can be an issue if they had to comprehend traffic lights in western countries. A summary of colors that countries perceive differently are shown below in Table 3.2.

The values of aesthetic symbols vary by culture, and some are preferred more than others by certain cultures (Alden, Steenkamp, and Batra, 1999). Icons and illustrations, too, are culture dependent. The symbol that means "all the good" in Hinduism can be mistaken to be the Nazi swastika symbol by Westerners (Lindholm et al., 2003). Semiotics has also been found useful to understand cultural differences in advertising (Alden et al., 1999). Functional and emotional needs, which are the main factors that influence use of technology, vary by cultures, too. What may be pleasurable or socially appropriate is dependent on culture (Lindholm, Keinonen,

TABLE 3.2
Summary of Colors Perceived Differently by Countries

Color	Countries That Find It Auspicious	Countries That Find It Unlucky
Red	China, India, Australian aborigines, United States, France	South Africa, Russia, Japan
Orange	Ireland, Netherlands	
Yellow	China, India, Japan	United States, France
Green	Ireland, Japan, Middle East	China, France
Blue	China, Iran, United States, France	Japan
White	United States, Europe	India, China, Japan,
Black		Thailand, United States, Europe

Source: Adapted from Kyrnin, J. (n.d.). Color Symbolism Chart by Culture. About.com. Retrieved from http://webdesign.about.com/od/color/a/bl_colorculture.htm; Marcus, A. (2001). International and intercultural user interfaces. In *User Interfaces for All: Concepts, Methods, and Tools* (Illustrate). Psychology Press; Barber, W., and Badre, A. (1998). Culturability: The merging of culture and usability. *Proceedings of the 4th Conference on Human Factors and the Web.* Retrieved on 1 November 2010 from http://www.research.att.com/conf/hfWeb/.

and Kiljander, 2003). Malaysians were found to react unfavorably to a product with green packaging, because green represents the jungle and all the diseases it brings forth (Ricks, Arpan, and Fu, 1974). On the other hand, green represents fertility in Egypt, safety in the United States, and criminality in France (Barber and Badre, 1998). Brides in western cultures often where white, whereas white is worn during funerals in India (Singh and Baack, 2004). With regards to car purchases, Chinese costumers were found to prefer white, whereas American consumers preferred the automobiles to be silver, black, or white (Johnson and Lieh-Ching, 2000).

While, the semiotics theory can offer a valuable perspective on the role that cultural differences play, there needs to be cultural research done in the realm of product design, outside of national shape or color (Lee, 2004).

In order to understand cultural product design, a cultural product design model was developed in 2006 (Lin, Sun, Chang, Chan, Hsieh, and Huang, 2007). This model consists of three main portions: a conceptual model, research method, and design process. The conceptual model deals with getting cultural features from objects and then transferring that to a model meant to design products. The research method includes three phases: identify, translate, and implement. The design process consists of four steps: investigate by setting a scenario, interact, by telling a story, develop, by writing a script, and finally, implement by designing a product. This model was used to analyze how Taiwanese aboriginal cultural elements have made their way into modern design (Lin, 2007). However, there is still a great need for more cohesive research to help designers develop culturally-acceptable products (Aykin 2005).

Similarly, activity theory has been found to be useful in explaining the impact of culture in the human–computer interaction domain (Honold, 2000). Activity theory is considered to have its roots in Soviet Union psychology in the early 1900s, but gained popularity toward the late 1980s (Nardi, 1996). In this theory, the unit of analysis is the activity of an individual. In other words, conscious, deliberate actions are not a group of different cognitive matters but a mere practice or an action. Each activity can be broken down into separate actions, and then each action is executed by processes that involve the goal and the context of the action. For instance, driving a car in the beginning consists of several conscious actions such as shifting gears, etc., but they eventually become automatic actions (Honold, 2000). Activity theory suggests that an action is performed due to the combined results of an individual interacting with the right tool as well as the environment (Nardi, 1996). Acknowledgment of the environment makes this theory extremely compatible with cross-cultural studies, since culture has been found to influence the way individuals think and act (Hofstede, 1982). This theory can be useful to understand the impact of culture on electronic products.

In the marketing domain, the Wheel of Consumer Analysis is a model used to understand consumer behavior (Peter and Olsen, 1994). According to this model, purchasing behavior depends on three aspects: environment, behavior, and affect/cognition. Affect/cognition is important to be understood because it deals with the internal reactions and attitudes that a consumer may have toward aspects of the external environment. Positive and negative attitudes have an impact on decisions. Behavior deals with what the customer actually does, and it is important that it is understood well. Finally, the environment deals with all the aspects of the external environment that the consumer is exposed to. Climate, geographical location, families, and friends are all part of the environment. This model has been used to understand cross-cultural consumer behavior (Johnson and Chang, 2000).

In the human-computer interaction domain, Triandis' model of subjective culture and social behavior (Triandis, 1994) was adapted to form a consumer behavior model (Lee, 2000). Triandis (1994) hypothesized that there were three main factors that have an impact on social behavior; subjective culture, experience, and the current situation. Subjective culture refers to the rules, roles, and categories in a culture and is supposed to influence outlooks toward products, consequences of purchasing, and habits through traditions and experience. Past experience refers to contemplation, usage, and purchase of the product and is supposed to influence outlooks toward products, consequences of purchasing, expectations based on referrals, and habits through traditions. Perception of the situation refers to the environment and facilitating conditions (Triandis, 1980) and is supposed to influence outlooks toward products, consequences of purchasing, and habits through traditions and experience. The consumer behavior model proposed by Lee (2000) continued to have subjective culture, experience, and situation as the main constructs, and hypothesized that *experience* had an influence on habit, while *subjective culture* had an influence on attitude, habit, consequences of purchase, purchase affect, self-definition, and referent expectations, and that these two constructs couple with *situation* had an influence on purchase intentions. This model was empirically tested with samples in Singapore, Korea, Hong Kong, and the United States, on a decision to buy a camera, and was found to explain the cultural influences on the purchase decision (Lee, 2000)

Finally, the consumer choice model was developed by Tybout and Hauser (1981) and is an extension of an older model by Hauser and Urban (1977). It starts with examining the environment and its physical characteristics that are observable. Physical characteristics are believed to influence perceptions and opinions. These characteristics even include observations of outside reactions to the product or reactions to marketing gimmicks (Tybout and Hauser, 1981). Thus, this model extends the consumer model put forth by Hauser and Urban (1977) in two ways: one, it distinguishes between characteristics of the environment and perceptions and then draws a relationship between the two. Second, it acknowledges the impact of the environment's physical characteristics on perceptions. Previously, only a product's physical characteristics were believed to impact the perception of a product. Choice of a product may be influenced by physical characteristics of a product, perceptions of the characteristics, and perception of the environment. Thus, this model suggests that consumer behavior is dependent on either the physical characteristics of the environment, or perceptions, or it could even be dependent on both (Tybout and Hauser, 1981), making it very applicable for use in the cultural domain.

RESEARCH METHODS

There are a number of research methods that can be used to understand the role of culture in the design and evaluation of consumer products. A summary of advantages and disadvantages of some of the methods, are shown in Table 3.3 below and then described in detail.

Surveys

While surveys are used for a variety of reasons, they usually have three main objectives: description, explanation, and exploration (Babbie, 1998). Surveys are usually conducted in two ways: questionnaires and interviews. Most often, the unit of analysis of a survey is a person, although it may not always be the case. For instance, in market research, the consumer sampled is the unit of analysis, but in group research, each group becomes the unit of analysis. In cross-sectional surveys, data are collected from a sample at a certain point in time to describe the behavior of the population at that same point in time. For instance, these surveys are used to determine purchasing intentions of consumers (Juster, 1966). In longitudinal studies, data are collected at various points in time. Longitudinal studies can be conducted in numerous ways. Trend studies include sampling and surveying a population at different points in time. While the population studied stays the same, different individuals of the population may be measured each time. Cohort studies on the other hand, pay attention to the same population at different points in time. For instance, a cohort study was used to understand whether vegetable and fruit consumption in the Netherlands was related to the risk of stomach cancer (Botterweck, van den Brandt, Goldbohm, 1998). Panel studies also collect data over time, but they collect the data from the same sample each time. While the main advantage of surveys is that it can be applied to any unit of analysis, results can also be misinterpreted and overgeneralized, making the researcher fall into the ecological fallacy trap (Robinson, 1950).

TABLE 3.3
Summary of Advantages and Disadvantages of Research Methods

Research Method	Advantages	Disadvantages
Questionnaire	Easy to administer Time effective Can produce direct quantitative data	Results can depend on how well the survey has been translated
Focus group	Can listen to multiple consumers at once, thus can save time	Can be difficult to gather different people together Can be difficult to analyze and transcribe
Interview	Can probe and gain better understanding More personal that questionnaires	Can be time consuming Can be dependent on style of interviewer
Observations	Less intrusive than interviews Can record sequence of events in the natural environment	Can be time consuming Can be expensive Participants may behave differently when they are aware they're being observed
Online research methods	Cost-effective Time effective Becoming exceedingly popular as web access increases	Limits sample to those who have Internet access Not yet useful for underdeveloped areas
Scanner data	Very useful in order to collect a lot of data Convenient to use	May produce too much information May be difficult to use in underdeveloped locations
Laboratory experiments	Useful to generate quantitative data Useful during prototype testing. Based on inputs from participant; new design can be generated	Can be expensive to simulate environment realistically Simulated environment will not include unanticipated occurrences

QUESTIONNAIRES

Questionnaires are extremely popular due to cost efficiency. However it must be acknowledged that the use of questionnaires can become expensive when they need to be translated to different languages. One study found that it was actually cheaper and more efficient to use telephonic interviews and mail surveys coupled with reminder calls, when surveying a group with limited-English proficiency (Metzger, Kaplan, Sorkin, Clarridge, and Phillips, 2004).

Online questionnaires are gaining increasing popularity because of time and cost conveniences. Currently, there are quite a few online tools that can be used to produce quite powerful surveys. For instance, an online questionnaire was used to understand perceptions of Gmail between users in India and the United States, however the researcher cautions that based on the difference between the quality and length of open-ended answers between the two samples, online tools might not be suitable for all cultures (Kothaneth, 2010).

INTERVIEWS

Surveys can also be conducted in the form of interviews. There are three main types of interviews conducted to collect information: structured, semistructured, and unstructured (Santiago, 2009). Structured interviews, as the name implies, stick to a very tight set of rules. The interview protocol, including the exact wording of the questions, is decided beforehand and deviation from the protocol is not permitted. Often, even the behavior and demeanor of the interviewer is decided and must be kept constant throughout all interviews. There should be very minimal reactions to the interviewee's responses. Semistructured interviews, on the other hand, are less rigid than structured ones. While there is a protocol developed beforehand, interviewers can deviate from it as they deem fit. This interview type is suggested when the topic is more personal to the interviewees. Semistructured interviews are well-designed for exploring opinions of respondents and allow prodding for more clarification. They are also perfect for groups who are from varied backgrounds (Barriball and While, 1994). Finally, an unstructured interview is the least rigid out of the three. The protocol usually consists of a basic checklist that serves as a guideline to make sure certain topics are covered. These are most efficient to gather a lot of information, and the interview often resembles a conversation (Santiago, 2009).

The idea of using interviews as the only means of gathering data has been critiqued because of the data can be confounded by the interviewees' emotional states at the time of the interview (Patton, 1980).

Semi-structured interviews seem to be very suitable for cross-cultural research due to their ability to explore opinions and handle a diverse set of participants (Barriball and While, 1994).

FOCUS GROUPS

Focus groups are useful to listen and conducting information (Kruger and Casey, 2009). It is a very effective way to get opinions from people about how they feel about a subject or a product. It is recommended to hold a focus group of 5 to 10 people, overseen by a facilitator. They are similar to personal interviews with the exception that they involve more people at a time. Focus groups are especially popular when conducting market research because they have been known to yield extremely useful data. They are very efficient to gain understanding, in testing prototypes, and evaluating a product (Kruger and Casey, 2009).

Research has also shown that the methodology of conducting a focus group must be executed differently in different cultures. For instance, a study of focus group interviews in East Asia revealed that it was important to build a rapport and trust between the participants of a focus group session. Lighthearted props, role-playing, and icebreakers were suggested to be employed to conduct efficient focus group interviews (Lee and Lee, 2009). Although focus groups are extremely useful in the domain of consumer research, it is recommended that it is not the only way to collect opinions (Calder, 1977).

Observations

The use of ethnographic research methods is extremely popular and have been applied frequently in fields that study people, practices, and culture. Data collection often occurs through participant observations. These are ideal in situations where there are a variety of different conditions, such as in cross-cultural research (Scheuch, 1989). For instance, Japanese consumer behavior was observed to understand why American products had low sales in Japan. It was found that Japanese consumers looked for products from major firms, thus popular brands specific to certain products were not impressive to the Japanese (Perner, 2010).

The use of ethnographic methods has been found to be useful to understand the use and application of technological systems (Macualay, Benyon, and Crerar, 2000). Supporters of this technique assert that designers will only be able to create realistic designs if they fully understand the use and context of the product that they are designing (Suchman, 1995). Many researchers take field notes to help them preserve the sequence of actions (Brandt, 1972). Care should be taken that the researcher does not intrude or get in the way of the consumer that is being observed.

ONLINE RESEARCH METHODS

In this new era of advanced web access, there is an increasing amount of web-based research methods. For instance, mail surveys are now being converted from paper-based questionnaires to Internet-based ones. Online surveys are especially useful because it can take full advantage of *conditional branching*, which basically refers to being able to jump to a question based on a type of response (Perner, 2010). Online research can also save a lot of time and money (Wright, 2005). The main limitation with using online research methods is that it limits the sample to those who have access to the Internet. Another disadvantage with using online research methods is that it is difficult to control. Multiple entries from the same person can be made possible by the use of multiple email addresses. However, newer online research packages offer response tracking devices (Wright, 2005). "Netnography" is a new technique that has been developed to understand the behavior of consumers in online communities. It is basically the adaptation of traditional ethnography methodologies to the online world (Kozinets, 2002). This technique was used to understand the buying behavior of sunscreen among Brazilian women (Figueiredo, 2011).

Scanner Data

As supermarkets get more competitive and consumers have more choices to choose from, it is becoming quite common for stores to offer free reward cards for consumers to use to avail discounts. These are easy to use and come in different sizes so they are easy to carry around as well. Scanner data are frequently used to make store-level and price-based customizations for consumers (Montgomery, 1997).

While they are definitely advantageous for the consumer, recent studies have shown that they can have an impact for consumer research as well Scanner checkouts in stores offer a lot of data, including details of products such as name, price, and quantity. Supplemental information that scanners provide includes customer counts and total sales (Jones, 1997). Retailers' scanner data are considered to be a valuable source of information that is often underutilized (Montgomery, 1997).

LABORATORY EXPERIMENTS

Laboratory experiments are post-positivist methods to obtain statistical data. They are simulations of the real-world environment, and common measures include observing participants interact with the product, measuring time taken to complete specific tasks, and testing performance. They are extremely useful when it comes to testing prototypes, but some researchers believe that the laboratory environment can never be realistic enough. Researchers who did a study on cultural usability and conducted experiments in three different countries recommended anticipating problems like differences in administrative procedures and differences in mental models of people from different backgrounds (Katre, Orngreen, and Yammiyavar, 2010). Perner (2010) has conducted a comprehensive compilation on consumer research methods and the reader is referred there for more details.

RECOMMENDATIONS FOR DESIGN OF CONSUMER PRODUCTS

UNIVERSAL DESIGN

Universal design has been defined as the method of designing products or surroundings that can be used by a variety of people, with different abilities, employing the products in different situations (Vanderheiden, 1997). Universal design is especially important when dealing with multiple countries with different cultural attributes. While a technology may be considered usable by some, it may not be to others, who in turn, will not adopt the technology or worse, criticize the technology openly, which could deter other users from adopting it. In order to overcome that hassle, one must try to implement "universal usability" or in other words, enhance the usability of an application so that almost anyone can use it (Schneiderman, 2000). Older technologies like telephones, televisions, etc., have reached their goal of universal usability for the most part, but technology that involves computers is still extremely difficult to employ for a lot of people (Schneiderman, 2000).

Universal design has become exceedingly popular because more and more researchers are starting to realize that culture and similar attributes do have an impact on using a product and what may be a good design to one may be a completely inappropriate design to others. There have been efforts to establish standards that can be used internationally. The International Standards Organization (ISO) is one such effort and according to them, standards enable fair trade between countries and facilitate innovation (ISO, 2010). Displays in Canada are now legally required to cater to both the English and the French speakers (Marcus, 2001).

Another similar initiative, Design for All (Dfa), is extremely popular in Europe because of its growing elderly and multicultural population. The European Commission widely promotes the Dfa to ensure better design. This initiative is founded on the

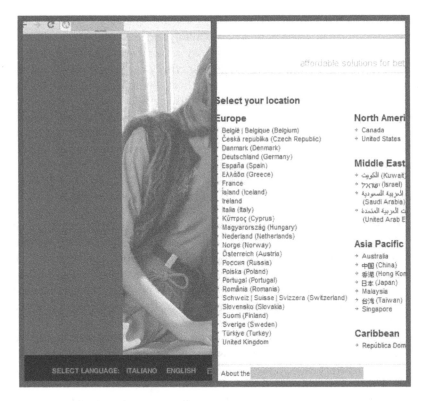

FIGURE 3.5   Example of websites that allow customization based on language and location.

principle that accessible, simple products improve the quality of life of users. (Europe's Information Society, 2008). The European Union has a number of initiatives to make sure that everyone is involved. These include e-Accessibility, Socio-Cultural eInclusion, and Geographical eInclusion (Europe's Information Society, 2008).

Research on this type of design has sparked the coinage of new words. *Internalization* refers to issues of geographic, political, and language type issues (Marcus, 2001). *Culturability* refers to the relationship between culture and user interface design (Barber and Badre, 1998). Websites have now started to cater to different cultural needs by including abilities to choose from different languages and culture-specific information to enhance the overall user experience (as shown in Figure 3.5).

DESIGN RECOMMENDATIONS BASED ON PRINCIPLES OF UNIVERSAL DESIGN

The principles of universal design were initially developed to make buildings and products more accessible for the both the able-bodied and those who were disabled physically. They are now becoming applied to the realm of cross-cultural design as well, since they provide rules and standards developing designs that can be used by people of all backgrounds.

All these design principles can be efficiently followed by making sure that people from different backgrounds and experiences are included in the design process as early as possible. Prototypes of products should be given to people at different locations and observations should be made as to how the prototypes are used in the outside world. Use of formative evaluation techniques early on in the process is extremely important to ensure good design. There are mainly two types of formative evaluation methods: analytical and empirical. Analytical methods can predict problems that users may encounter. They include methods like heuristic evaluations and cognitive walkthroughs. Empirical methods, on the other hand, involve users and are also called *usability testing methods*. They involve think-aloud laboratory-based tasks (Capra, 2006). Scenarios are another method for usability evaluation. Formative evaluations require mainly qualitative data, such as interviews and critical incident descriptions (Capra, 2006). In particular, the principles of universal design should be followed, but with a focus on the cross-cultural domain, as explained in the text that follows.

1. **Design for equitable use**. This principle ensures inclusiveness, and that all groups are to be kept in mind, while creating a design. Literacy and linguistic abilities should also be recognized when creating a universal design. Labels and graphics should be easy to understand and should not be offensive to anyone. For instance, Apple's iPhones now have a multilanguage keyboard, which enables people with different linguistic abilities to use the phone with ease (as shown in Figure 3.6).
2. **Flexibility in use.** This principle ensures that products can be customizable to a certain extent since different users may have different preferences. This allows different users to use products in a manner that is most comfortable to them.
3. **Simple and intuitive**. Every product should be easy to use, and if it makes use of graphics, the graphics must be universally intuitive. This can be a challenge with the vast range of consumer preferences, but understanding different consumer requirements before starting the design process will be helpful. Providing affordances, which "suggest" to the user how to use the

FIGURE 3.6　Example of mobile phone keyboard that is designed for equitable use.

FIGURE 3.7   Examples of common worldwide symbols that are used for electronic items.

product, will be extremely useful (Norman, 1998). Other theories such as the Gibson's theory on perception asserted that affordances are key for efficient interaction (Gibson, 1979). Good design will make use of affordances that enable understanding and using the product effectively. Affordances can be of many types, including perceptual, movement, and shape affordances. Symbols can also act as affordances. Examples of symbols that are used worldwide for electronic items are shown in Figure 3.7.

4. **Perceptible information.** This principle suggests that information should be provided in multiple ways so that anyone has access to it, no matter what their background or experience is. For instance, voice menus in the United States are becoming more accessible to Spanish-speaking people, by offering options to listen to the menu in Spanish.

5. **Tolerance for error.** This principle suggests that there should be minimal consequences for making mistakes. Thus, if a user presses a wrong button due to inexperience, he should be able to go back and restart without much penalty. Prompts asking the user to confirm if that's the intended action are extremely useful. Gmail®, an extremely popular electronic mail provider, recently started providing useful alerts if it appears the user is making a mistake. Figure 3.8 following shows a prompt for when the user appears to have forgotten to attach a file, as mentioned in the body of the email.

6. **Low physical/mental effort.** This principle suggests that a product should be designed such that it induces low fatigue and requires little effort. This is especially crucial for consumer products that are sold globally. Ideally, features should be easy and intuitive to use, so that users don't need to waste too much time trying to figure it out. This is also important for

FIGURE 3.8   Example of email design that helps the user avoid mistakes.

FIGURE 3.9 Example of a coffee maker with confusing icons and multiple buttons.

safety. An example of a design that induces high mental effort is shown in Figure 3.9.

7. **Size and space for approach and use.** This principle suggests that a product should be designed to enable use by individuals of different body sizes, types, and abilities. For instance, while smartphones provide the owner with anytime access to vast amounts of information, it has also exponentially grown in size with reports indicating that mobile phones are too big to carry around in pockets anymore (Sutter, 2010).

GENERAL RECOMMENDATIONS

In this vast and complex domain where design meets culture, it is important the governments across the world follow the European Union's example and contribute and adhere to initiatives like Design for All. Unless, there are inputs from people of different backgrounds, abilities, and experiences, it will be extremely difficult to develop standards for universal design. Standards that already exist must be rechecked because they might not have been developed keeping customers across different cultures in mind. For instance, staff at an African port threw boxes of wine glasses away because they mistook the symbol for 'fragile' (popularly depicted by a broken wine glass) for broken wine glasses ("Cross Cultural Marketing Blunders", n.d.). Since consumer products encompass such a range of devices and tools, it is safe to say that investing in good standards will be worth it and can be used for a long while. However, it is important not to underestimate the power of acculturation that stems from the throngs of people migrating from one country to the next. Migration between countries is now becoming extremely common and people are starting to get adjusted or "accultured" to the new land. *Acculturation* refers to the impact of another cultural group on one's behavior, values, and beliefs (Berry, 1980). It is usually considered to be of two types: old behaviors are replaced by new ones or new behaviors are acquired without replacing the old ones. Changes in food preference, clothing, and ethnic identity are all attributed to acculturation (Berry, 1980). Recent studies have found that culture can be primed in bicultural individuals through exposure to languages, and cultural icons, suggesting that acculturation may be a dynamic construct (Lechuga, 2008). Standards must be revisited from time to time as people's needs and abilities might have changed.

As more marketers realize that qualitative methods are crucial to understand consumer products, it is important to acknowledge that a mix of quantitative and qualitative methods will yield the highest validity (Calder, 1977). Thus, it is recommended to use a more pragmatic approach and to use mixed methods for data collection and analysis. It is extremely vital that you understand the culture of the population that you want to study before deciding what type of research method is suitable. For instance, Chavan (2005) asserted that traditional usability testing methods were not suitable for Indian users since Indians were hesitant to criticize products and often work around problems.

There should be ample customer support and enough advertising to attract consumers from nations who tend to be hesitant to try new products. Consumers from allocentric cultures have been found to gain experience and knowledge about a product by seeking references from friends and family as well as reading about products in specialty magazines (Bettman, Johnson, and Payne, 1991). It is also recommended that the public relations representative is well versed on the culture and does not say or do anything offensive. Seemingly innocent comments can hurt feelings and lead to boycotting of products in extreme cases.

Since it has been acknowledged that culture adds a dimension of complexity to any domain, it is important to pay heed to the entire situation because consumers can be influenced by a lot of things. Consumer influences have been previously categorized into five types: physical surroundings, social environment, time-based perspective, type of task, and previous states (Belk, 1975). A good rule of thumb to follow is to get acquainted with their likes and dislikes and to run all ideas by someone well versed in that particular culture's customs before publicizing the product. It is extremely important that nothing about the product is offensive to the culture in any way. Needless to say, it is extremely important that nothing about the product itself is offensive to the culture in any way. For instance, even though Japan's cars typically have steering wheels on the right side, affluent Japanese were found to prefer the steering wheel on the left (which is clearly indicative of a U.S. manufactured car) because they perceive it as a status symbol (Taylor, 1983). For instance, the popular pen company, Mont Blanc had to suspend sales and issue a public apology after millions of people were offended by the "Gandhi pen" (Nelson, 2010). The most efficient way to ensure this is involving people who represent that culture in all steps of the design process.

Since this is a relatively new, but increasingly popular domain of research, designers should pay attention to the recent advents in research and pay heed to guidelines and recommendations. For instance, Aykin and Milewski (2005) suggest over fifty guidelines to enhance web page design from a cross-cultural perspective. Some of their guidelines suggest avoiding acronyms, jokes and idioms, and to reduce the use of colloquial language. For more details, the reader is referred to their original text.

Designers and marketers should enter the domain with an open mind. Consumer product designers should try and accept the ways of different parts of the world and conform to those specifications. While some countries will be open to trying the ways of affluent countries, others are very keen to stick to their traditional methods. These methods may be archaic to the inexperienced, but it is advised to design and market products keeping their traditions in mind. Most importantly, it is recommended to have a very diverse team of designers and human factors engineers take part in the design exercise. Clearly assigned goals will foster job responsibility and reduce duplication of work. Diversity in teams brings in different perspectives, which is extremely important when designing a culturally acceptable product. Finally, iterative user-testing at different stages of research, with representative users is imperative to develop an acceptable and successful product.

REFERENCES

Adler, N. (1991). International dimensions of organizational behavior. 2nd Ed. Boston, Kent.
Ajzen, I. (1985). From intentions to actions: A theory of planned behavior. In J. Kuhi and J.
Ajzen, I., and Fishbein, M. (1980). *Understanding attitudes and predicting social behavior.* Englewood Cliffs, NJ, Prentice-Hall.
Alden, D., Steenkamp, J.-B., and Batra, R. (1999). Brand Positioning Through Advertising in Asia, North America and Europe: the Role of Global Consumer Culture. *Journal of Marketing,* 63(1), 75–87.
Aykin, N. (2005). Overview: where to start and what to consider. In *Usability and internationalization of information technology* (pp. 3–20). Mahwah, NJ, Lawrence Erlbaum Associates Publishers
Aykin, N. and Milewski, A. E. (2005). Practical issues and guidelines for international information display. In *Usability and internationalization of Information Technology* (pp. 21–50). Mahwah, NJ: Lawrence Erlbaum Associates Publishers.
Babbie, E. (1998). Survey Research Methods. Belmont, CA, Wadsworth, Inc. 2nd ed., pp. 51–63.
Barber, W., and Badre, A. (1998.) Culturability: The merging of culture and usability. *Proceedings of the 4th Conference on Human Factors and the Web.* Basking Ridge, NJ, U.S.A. Retrieved November 2010 from http://research.microsoft.com/enus/um/people/marycz/hfweb98/barber/index.htm
Becker C., (1997). Hair and Cosmetic Products in the Japanese Market. *Marketing and Research Today,* pp. 31–36.
Becker, G. S. (1965). A theory of the allocation of time. *Economic Journal,* 75 (September), 493–517.
Belk, R., and Pollay, R. (1985). Materialism and Magazine Advertising during the Twentieth Century. In *Advances in Consumer Research.* Vol. 12. Association for Consumer Research, pp. 394–398.
Belk, R. W. (1975). Situational variables and consumer behavior. *Journal of Consumer Research,* 2, 157–163.
Berry, J. W. (1980). Acculturation as varieties of adaptation. In *Acculturation: Theory, models and some new findings.* Boulder, CO, Westview.

Bettman, J. R., Johnson, E. J., and Payne, J. W. (1991). Consumer decision making. In *Handbook of consumer behavior.* Englewood Cliffs, NJ, Prentice Hall.

Bond, M. H. (1983). How language variation affects inter-cultural differentiation of values by Hong Kong bilinguals. *Journal of Language and Social Psychology,* 2(1), 57–67.

Botterweck, A., Brandt, P. a van den, and Goldbohm, R. (1998). A prospective cohort study on vegetable and fruit consumption and stomach cancer risk in The Netherlands. *American journal of epidemiology,* 148(9), 842–53. Retrieved from http://www.ncbi.nlm.nih.gov/pubmed/9801014.

Brandt, R.M. (1972). *Studying Behavior in Natural Settings,* New York.

Calder, B. J. (1977). Focus Groups and the Nature of Qualitative Marketing Research. *Journal of Marketing Research,* 14(3), 353–364.

Capra, M. (2006). *Usability Problem Description and the Evaluator Effect in Usability Testing.* Blacksburg, Virginia, Virginia Polytechnic Institute.

Chavan, A. L. (2005). Another culture, another method. In *Proceedings of the 11th International Conference on Human-Computer Interaction.* Mahwah, NJ, Lawrence Erlbaum Associates.

Cross Cultural Marketing Blunders (n.d), In *Kwintessential.* Retrieved from http://www.kwintessential.co.uk/cultural-services/articles/crosscultural-marketing.html

Daghfous, N., Petrof, J. V., and Pons, F. (1999). Values and adoption of innovations: A cross-cultural study. *Journal of Consumer Marketing,* 16(4), 314–329.

Danziger, P. M. (2005). *Let them eat cake. Marketing luxury to the masses—as well as the classes.* Chicago, Dearborn Trade Publishing

de Mooij, K. M. (2004). *Consumer behavior and culture: consequences for global marketing and advertising* (Illustrate, pp. 139–145). Sage Publishing, CA.

de Mooij, M. (2000). The future is predictable for international marketers: Converging incomes lead to diverging consumer behavior. *International Marketing Review,* Vol. 17., Issue 2, pp. 103–113.

de Mooij, M., and Hofstede G. (2002). Convergence and divergence in consumer behavior: implications for international retailing. *Journal of Retailing,* Vol. 78, Issue 1. pp. 61–69.

Depeux N. (n.d.). Semiotic Analysis and China's Bottled Water Market. Green Book. Retrieved June 1st, 2012, from http://www.greenbook.org/marketing-research.cfm/semiotic-analysis-and-chinas-bottled-water-market

Douglas M., and Isherwood, I. (1978). *The World of Goods: Towards an Anthropology of Consumption,* New York.

Douglas, M. (1966). *Purity and Danger: An Analysis of Concepts of Pollution and Taboo,* Harmondsworth, England, Penguin Books.

Epstein, Z. (2008). Nokia Releases Q4 2007 Figures; Global Sales up, US Sales Down. Boy Genius Report. Retrieved from http://www.boygeniusreport.com/2008/01/24/nokia-releases-q4–2007-figures-global-sales-up-us-sales-down/.

Europe's Information Society Thematic Portal (2008). Report on Policy and Design for All (DfA) (2008). Retrieved 1st October, 2010 from http://www.epractice.eu/en/library/281650

Evers, V. and Day, D. (1997). *The Role of Culture in Interface Acceptance.* In Proceedings of Human Computer Interaction, Interact '97, pp. 260–267. Chapman & Hall, London.

Figueiredo D. (2011). Sunscreen Culture among Brazilian Women: A Netnographic Approach. Message posted to http://www.netnography.com/showthread.php?5261-Sunscreen-Culture-Among-Brazilian-Women-a-Netnographic-Approach.

Fischer, E., and Arnold S. (1994). Sex, Gender Identity, Gender Role Attitudes, and Consumer Behavior. *Psychology and Marketing,* 11(2), 163–182.

Frith, K., Shaw, P., and Cheng, H. (2005). The Construction of Beauty: A Cross-Cultural Analysis of Women's Magazine Advertising. *International Communication Association* (March), 56–70.

Gengenbach, H. (2003). Boundaries of Beauty: Tattooed Secrets of Women's History in Magude District, Southern Mozambique. *Journal of Women's History*, 14(4), 106–141.

Geser, H. (2004). *Towards a sociological theory of the mobile phone*. University of Zurich. Retrieved February, 14, 2005. Retrieved from http://www.itu.dk/people/ldn/Lars/Geser_lektion6.pdf.

Gibson, J.J. (1979). *The Ecological Approach to Visual Perception*. Boston, Houghton Mifflin.

Goldman, D. (1999). Paradox of pleasures. *American Demographics*, 21(5), ProQuest Database.

Green R., Cunningham C., and Cunningham W. (1974). Cross Cultural Consumer Profiles: An Exploratory Investigation. In *Advances in Consumer Research*. Volume 01, pp. 136–144.

Hall, E. T. (1959). *The Silent Language*. Garden City, NY, Doubleday.

Hampden-Turner, C. and Trompenaars, A. (1997). The Seven Cultures of Capitalism: Value System for Creating Wealth in the United States, Britain, Japan, Germany, France, Sweden and the Netherlands. Piatkus, London, UK.

Hangen, T. J. (2001). *Redeeming the dial: Radio, religion, and popular culture in America*. Chapel Hill, NC, University of North Carolina Press.

Hauck, W. E., and Stanforth, N. (2007). Cohort perception of luxury goods and services. *Journal of Fashion Marketing and Management*, 11(2), 175–188.

Hauser, J. and Urban, G. (1977). A Normative Methodology for Modeling Consumer Response to Innovation. *Operations Research*, 25(4), 579–619.

Hempel, J. (2009). Nokia's North America problem. *Fortune Magazine*. Retrieved from http://money.cnn.com/2009/01/12/technology/hempel_nokia.fortune/index.htm.

Hill, R. L. (1970). Family development in three generations. Cambridge, Massachusetts: Schenkman.

Hofstede, G. (1982). Dimensions of national cultures. *Diversity and unity in cross-cultural psychology*. Lisse, Netherlands, Swets and Zeitlinger.

Hofstede, G. (1984). *Culture's consequences: International differences in work-related values*. Sage Publications. pp. 77–99.

Hofstede, G. (2003). *Culture's Consequences, Comparing Values, Behaviors, Institutions, and Organizations across Nation*. Sage Publications, Second Edition.

Holman, R. (1976). Communicational Properties of Women's Clothing: Isolation of Discriminable Clothing Ensembles and Identification of Attributions Made to One Person Wearing Each Ensemble. Unpublished PhD dissertation, the University of Texas at Austin.

Honold, P. (2000). Culture and Context: An Empirical Study for the Development of a Framework for the Elicitation of Cultural Influence in Product Usage. *International Journal of Human-Computer Interaction*, 12(3&4), 327–345.

Iyer, G., Laplaca, P., and Sharma, A. (2006). Innovation and new product introductions in emerging markets: Strategic recommendations for the Indian market. *Industrial Marketing Management*, 35(3), 373–382.

Johnson, W. C. and Chang, L. (2000). *A Comparison of Car Buying Behavior Between American and Chinese People Living in North America: An Exploratory Study*. Presented at Southwestern Marketing Association Conference, San Antonio, TX.

Jones, E. (1997). An Analysis of Consumer Food Shopping Behavior Using Supermarket Scanner Data: Differences by Income and Location. *American Journal of Agricultural Economics*, 79(5), 1437–1443.

Juster, T. (1966). Consumer Buying Intentions and Purchase Probability: An Experiment in Survey Design. *Journal of American Statistical Association*, 61(315), 658–696. Retrieved from http://www.jstor.org/stable/2282779.

Kahle, Lynn R. (1986). The Nine Nations of North America and the Value Basis of Geographic Segmentation. *Journal of Marketing*, 50(2) April.

Katre, D., Orngreen, R., and Yammiyavar, P. (2010). Human Work Interaction Design: Usability in Social, Cultural and Organizational Contexts. Springer-Verlag, New York, 1st ed.

Klein, H. (2004). Cognition in Natural Settings: The Cultural Lens Model. Advances in Human Performance and Cognitive, 4(4), 249–280. Retrieved from http://www.emeraldinsight.com/books.htm?chapterid=1761993&show=pdf.

Kluckhohn, F. and Strodtbeck, F. L. (1961). Variations in value orientations. Evanston, IL. Row, Peterson.

Kothaneth, S. (2010). A Pilot Study on the Cross-Cultural Acceptance of Technology. In *The Proceedings of the Human Factors and Ergonomics Society Annual Meeting* September 2010, 54: 1951–1955.

Kozinets, R. V. (2002). The Field behind the Screen: Using Netnography for Marketing Research in Online Communities. *Journal of Marketing Research,* 39(1), 61–72.

Kozinets, R. V. (2010). *Netnography: Doing Ethnographic Research Online.* London: Sage.

Krish, V. (2010, April 4). Nokia India has 54.1 percent market share, over 100 million phones sold in 2009 in India. *Fone Arena.* Retrieved from http://www.fonearena.com/blog/14770/nokia-india-has-54–1-percent-market-share-over-100-million-phones-sold-in-2009-in-india.html.

Krueger, R. and Casey, A. M. (2009). Focus groups: a practical guide for applied research (4th ed) Thousand Oaks, CA, Sage.

Kyrnin, J. (n.d.). Color Symbolism Chart by Culture. *About.com.* Retrieved from http://web-design.about.com/od/color/a/bl_colorculture.htm.

Lamb, C., Hair, J., and McDaniel, C. (2008). *Essentials of Marketing* (10th ed., pp. 301–303). Cengage Learning. Mason, OH, US.

Lane, H. and DiStefano, J. (1992). International management behavior: From policy to practice. Boston, PWS-Kent.

Lechuga, J. (2008). Is acculturation a dynamic construct? The influence of priming culture on acculturation. *Hispanic Journal of Behavioral Sciences,* 30, 324–339.

Lee, J. (2004). Design Methods for Cross-cultural Collaborative Design Project in Redmond, Durling and De Bono (Eds.) Design Research Society international conference: Futureground, Melbourne, Australia, November 17–21.

Lee, J. (2000). Adapting Triandis's Model of Subjective Culture and Social Behavior Relations to Consumer Behavior. *Journal of Consumer Psychology,* 9(2), 117–126.

Lee, J. J. and Lee, K. P. (2009). Facilitating dynamics of focus group interviews in East Asia: Evidence and tools by cross-cultural study. *International Journal of Design,* 3(1), 17–28.

Lerner, G. (1986). *The Creation of Patriarchy.* New York, Oxford University Press.

Liang, Y. F. (2005). The value of luxury brands. United Daily Newspaper.

Lighting up. Who likes a smoke? (2007, June 2nd). *The Economist.* Retrieved from http://www.economist.com/node/13766483.

Lin, R. T. (2007). Transforming Taiwan aboriginal cultural features into modern product design: A case study of a cross- cultural product design model. *International Journal of Design,* 1(2), 45–53.

Lin, R., Sun, M., Chang, Y., Chan, Y., Hsieh, Y., and Huang, Y. (2007). Designing "Culture" into Modern Product: A Case Study of Cultural Product Design. *Culture,* 4559, 146–153. Springer.

Lindholm, C., Keinonen, T., and Kiljander, H. (2003). Mobile usability: How Nokia Changed the Face of the Mobile Phone (pp. 97–103). McGraw-Hill, New York.

Macaulay, C., Benyon, D., and Crerar, A. (2000). Ethnography, theory and systems design: from intuition to insight. *Human-Computer Studies,* 53, 35–60.

Marcus, A. (2001). International and Intercultural User Interfaces. In *User interfaces for all: concepts, methods, and tools* (Illustrate). Psychology Press. Mahwah, NJ.

Marcus, A. (2002). User-interface design, culture, and the future. In *Proceedings of the Working Conference on Advanced Visual Interfaces—AVI '02)*, New York.

Mccracken, G. (1986). Culture and Consumption: A Theoretical Account of the Structure and Movement of the Cultural Meaning of Consumer Goods. *The Journal of Consumer Research,* 13(1), 71–84.

Mick, D. G. (1986). Consumer Research and Semiotics: of Exploring the Morphology of Signs, Symbols and Significance. *The Journal of Consumer Research,* 13(2), 196–213.

Minkler, L. and Cosgel, M. (2004). Economics Working Papers. Paper 200403. http://digitalcommons.uconn.edu/econ_wpapers/200403.

Montgomery, A. L. (1997). Creating Micro-Marketing Pricing Strategies Using Supermarket Scanner Data. *Marketing Science,* 16(4), 315–337.

Moon, J., Chadee, D., and Tikoo, S. (2008). Culture, product type, and price influences on consumer purchase intention to buy personalized products online. *Journal of Business Research,* 61(1), 31–39.

Munson, J. M. and McIntyre, S. (1979). Developing Practical Procedures for the Measurement of Personal Values in Cross-Cultural Marketing. *Journal of Marketing Research,* 16, 55–60.

Nardi, B. (1996). Concepts of Cognition and Consciousness: Four Voices. *Australian Journal of Information Systems,* 4(1), 64–79.

Neelankavil, J. P., Mathur, A., and Zhang, Y. (2000). Determinants of managerial performance: A cross-cultural comparison of the perceptions of middle-level managers in four countries. *Journal of International Business Studies,* 31(1), 121–140.

Nelson, D. (2010). Mont Blanc apologizes for Gandhi pen. *The Telegraph.* Retrieved from http://www.telegraph.co.uk/news/worldnews/asia/india/7316148/Mont-Blanc-apologises-for-Gandhi-pen.html.

Ngo-Metzger, Q., Kaplan, S. H., Sorkin, D. H., Clarridge, B. R., and Phillips, R. S. (2004). Surveying Minorities with Limited-English Proficiency. *Medical Care,* 42(9), 893–900.

Nisbett, R. E. (2003). *The Geography of Thought: How Asians and Westerners Think Differently and Why,* Brealey, London, 2003.

Norman, D. (1998). *The Design of Everyday Things,* London: MIT Press.

Number of Cell Phones Worldwide Hits 4.6B. (2010, February 18th). *CBS News.* Retrieved from http://www.cbsnews.com/stories/2010/02/15/business/main6209772.shtml.

O'Brien, J. K. (2009). Nokia Tries to Undo Blunders in U.S. *The New York Times.* Retrieved from http://www.nytimes.com/2009/10/19/technology/companies/19nokia.html?_r=1.

Oliver, J. D. and Lee, S.-H. (2010). Hybrid car purchase intentions: a cross-cultural analysis. *Journal of Consumer Marketing,* 27(2), 96–103. doi: 10.1108/07363761011027204.

Olson, J. C. and Peter, J. P. (1994). Understanding Consumer Behavior. Burr Ridge, IL: Richard D. Irwin, Inc.

Palan, K., Areni C., and Kiecker, P. (2001). Gender Role Incongruency and Memorable Gift Exchange Experiences. *Advances in Consumer Research,* 28. Eds. Mary C. Gilly and Joan Meyers-Levy. Provo, UT: Association for Consumer Research, in press.

Patton, M. (1980). *Qualitative evaluation methods,* Beverly Hills, CA, Sage.

Perner L. (2010). Consumer Behavior: The Psychology Of Marketing. In *USC Marshall.* Retrieved 13th November, 2011, from http://www.consumerpsychologist.com/.

Ricks, D. R., Arpan, J. S., and Fu, M. Y. (1974). Pitfalls in advertising overseas. *Journal of Advertising Research,* 14(6), 47–51.

Robinson W. S. (1950). Ecological Correlations and the Behavior and Individuals. *American Sociological Review,* pp. 351–357.

Rogers, Everett M. (1995). Diffusion of Innovations, Free Press, New York, NY.

Rokeach, M. (1973). *The Nature of Human Values,* New York, Free Press.

Santiago, N. (2009). Interview types: Structured, semi-structured, and unstructured. *San Jose Scholarly Research.* Retrieved from http://www.examiner.com/scholarly-research-in-san-jose/interview-types-structured-semi-structured-and-unstructured.

Schein (1999). *The Corporate Culture Survival Guide,* Jossey Bass, San Francisco, CA.

Scheuch, E. K. (1989). Theoretical Implications of Comparative Survey Research: Why the Wheel of Cross-Cultural Methodology Keeps on Being Reinvented. *International Sociology,* 4(2), 147–167.

Schneiderman, B. (2000). Universal Usability. *Communications of the ACM,* 43(5), May, pp. 84–91.

Scholderer, J. and Grunert, K. (2005). Consumers, food and convenience: The long way from resource constraints to actual consumption patterns. *Journal of Economic Psychology,* 26(1), 105–128. doi: 10.1016/j.joep.2002.08.001.

Senauer, B. (2001). *The food consumer in the 21st century: New research perspectives.* University of Minnesota. The Food Industry Center.

Sherry, J. and Camargo, E. (1987). May Your Life Be Marvelous: English Language Labelling Labeling and the Semiotics of Japanese Promotion. *Journal of Consumer Research,* 14 (September), 174–88.

Sheth, Newman, and Gross. (1991). Why We Buy What We Buy: A Theory of Consumption Value. *Journal of Business Research,* 22(2), pp. 159–170.

Shome, R. and Hegde, R. (2002). Culture, communication, and the challenge of globalization. *Critical Studies in Media Communication,* 19(2), 172–189. doi: 10.1080/07393180216560.

Singh N. and Baack, W. (2004). Web Site Adaptation: A Cross-Cultural Comparison of U.S. and Mexican Web Sites. *Journal of Computer-Mediated Communication,* Vol. 9 (4).

Spencer-Oatey (2000). *Culturally Speaking: Managing Rapport through Talk across Cultures,* Continuum International Publishing Group. London, UK.

Stephan (2004). An Overview of Intercultural Research: The Current State of Knowledge, CEE Publishing, London, UK.

Strauss, W. and Howe, N. (1991). *Generations: the History of America's Future,* New York, Morrow.

Suchman, L. (1995). Making work visible. *Communications of the ACM,* 38, 56–65.

Suh, T. and Kwon, I. (2002). Globalization and reluctant buyers. *International Market Review,* 19(6), 663–80.

Sutter, J. (2010). When phones are too big for pockets. *CNN Tech.* Retrieved from http://articles.cnn.com/2010–07–29/tech/5.inch.dell.streak_1_mobile-phone-htc-evo-dell-streak?_s=PM:TECH.

Taylor, D. (1992). *Global Software: Developing Applications for the International Market.* Springer Publishers.

Triandis, H. C. (1980). Values, attitudes and interpersonal behavior. *Nebraska Symposium on Motivation,* 27, 195–260.

Triandis, H. C. (1994). *Culture and social behavior,* McGraw-Hill, New York.

Triandis, H., Bontempo, R., Villareal, M., Asai, M., and Lucca, N. (1988). Individualism and Collectivism: Cross-Cultural Perspectives on Self-Ingroup Relationships. *Journal of Personality and Social Psychology,* 54, 323–338.

Tripathy, D. (2009). Nokia sees sales growth in India despite drought. *Reuters.* Retrieved from http://www.reuters.com/article/idUSDEL26706520090819.

Tse, D., Wong J., and Tan C. (1988). Towards Some Standardized Cross-Cultural Consumption Values. In *Advances in Consumer Research.* Volume 15, pp. 387–395.

Tybout, A. and Hauser, J. (1981). A Marketing Audit Using a Conceptual Model of Consumer Behavior: Application and Evaluation. *Journal of Marketing,* 45 (Summer), 82–101, 37–47.

Vanderheiden, G. C. (1997). Anywhere, Anytime (+ Anyone) Access to the Next Generation WWW. *Computer Networks and ISDN Systems,* 29, 1439–1446.

Vatikiotis, M. (1996). Children of plenty. *Far East Economic Review,* 159(49), 54–69

Wallace, S. and Yu, H.-C. (2009). The Effect of Culture on Usability: Comparing the Perceptions and Performance of Taiwanese and North American MP3 Player Users. *Journal of Usability Studies,* 4(3), 136–146.

Why Standards Matter (2010). *International Organization for Standardization (ISO).* Retrieved from http://www.iso.org/iso/about/discover-iso_why-standards-matter.htm.

Woodruff, A., Augustin, S., and Foucault, B. E. (2007). Sabbath Day Home Automation: It's like mixing technology and religion. *CHI 2007,* 527–536.

Wright, K. B. (2005). Researching Internet-based populations: Advantages and disadvantages of online survey research, online questionnaire authoring software packages, and web survey services. *Journal of Computer-Mediated Communication,* 10(3), article 11. http://jcmc.indiana.edu/vol10/issue3/wright.html

Wyche, S. P. and Grinter, R.E. (2009). Extraordinary Computing: Using Religion as a Lens for Reconsidering the Home. Proceedings of the ACM SIGCHI Conference on Human Factors in Computing Systems (CHI '09), Bostom, MA, Pgs. 749–758.

4 Human–Computer Systems

Kayenda T. Johnson

CONTENTS

Many cultures around the world work to harness both the social and financial benefits tied to the Internet and the international marketplace that the Internet undergirds. Opportunities are consistently present for businesses and organizations of all kinds to establish and expand their markets into other geographic locations. The key element to conducting business across geographic borders and engaging domestic and international marketplaces is a computer system that will bare a computer interface designed as the point of interaction for the end user. As businesses and organizations alike seize opportunities to market their products or services, it is essential for them to recognize that cultural differences among their potential user groups must be factored into the design of the computer interface. Within cultures people share common languages, values, beliefs, rituals, rules, and normative practices that comprise the way in which their society functions (Veroff and Goldberg, 1995). Consequently, the unique cultural attributes of the intended user groups must be considered for the design and development of human–computer systems, in order to mitigate the negative consequences produced when deviating from the users' mental models.

It is paramount that designers and developers of these human–computer systems, namely, computer interfaces, apply an understanding of the sociocultural contexts of use associated with the web-based and software products they are developing. Interface designs for computing systems that are not culturally appropriate, will run the risk of excluding some people groups, populations, regions, or countries from effective and safe user experiences with the products, services, or information being provided.

There are numerous theories associated with cultural differences, spanning academics areas such as anthropology, social psychology, human factors, human–computer interaction, and others. However, there is much less empiricism demonstrating the practical manner in which designers and developers can translate these theories into interface design elements for web-based design and software development. So, one could ask, what tools can designers and developers use to design within the sociocultural context of their targeted user groups? This chapter seeks to respond to that question by presenting a user-centered design framework that incorporates a variety of practical tools and techniques to facilitate cultural relevance in design.

GLOBALIZATION AND LOCALIZATION

Computer interfaces are everywhere. They drive business, education, government, and service industries nationally and internationally. Computer interfaces have become essential gateways for people to achieve their daily and common activities. For example, citizens of various countries and nations use interfaces within cell phones, kiosks, home appliances, industrial machines, and computers to communicate, rent movies, shop, play games, manage bank accounts, make health-related appointments, control production and assembly of products, cook, order pizza, conduct business, write letters, read books, and so on. Their ubiquitous nature provide endless opportunities to start and grow businesses, enhance and extend education, aid the organization and implementation of government, and expand recreation and entertainment into markets across geographic borders and beyond. In our global society, almost every aspect of living requires some level of interaction with computer interfaces.

Accounting for cultural variation among targeted end-user groups for international markets can take on one of two approaches: globalization and localization. Globalization, also known as internationalization, "refers to a broad range of processes necessary to prepare and launch products and company activities internationally (http://www.gala-global.org/view/terminology)." Localization describes the process

> … of adapting a product to a specific international language or culture so that it seems natural to that particular region, which includes translation, but goes much father. True localization considers language, culture, customs, technical and other characteristics of the target locale. While it frequently involves changes to the software writing system, it may also change the keyboard usage, fonts, date, time and monetary formats. Graphics, colors and sound effects also need to be culturally appropriate (http://www.gala-global.org/view/terminology).

CULTURAL MISMATCHES IN INTERFACE DESIGN

Attempts to determine what credence should be assigned to this notion of designing toward globalization or localization, may lead designers and business owners in international markets to ask, "Will cultural factors really affect the utility of my products and/or services? Will the end user really care about cultural details? Could the lack of cultural consideration really decrease the overall success and safety of my products? Do we really need to study and understand the sociocultural context of our target user groups?" The answer to these questions is yes. Products and systems, in general, can produce dangerous situations when they deviate from the user's mental models or ways of thinking. These deviations from the users' way of thinking can cause the users to experience negative consequences from frustration to severe physical injury or even death. The following discussion about the use of interface metaphors across geographic and cultural borders provides an example of the importance of cultural considerations within design. The implications of not considering cultural factors in this example will not lead to physical injury or death per se, but could indeed result in severe annoyance and frustration with the computer interface and the application it represents.

Some people believe in the universal relevance of the "desktop" metaphor for office working environments. Others contend that the desktop metaphor, which is influenced by mainstream American culture, is not fully applicable outside the borders of the United States. For example, Duncker (2002) discusses computing metaphors and emphasizes that while the general concept of a desktop metaphor transfers across cultures, the more granular aspects of the metaphor do not transfer. In American culture, folders are made from firm pieces of paper and have a tab for labeling; these folders are stored horizontally in filling cabinets and drawers. In other cultures like those of Japanese and European countries, folders are handled and stored differently. For instance, Japanese and European cultures store their files in lever arch files, which look like cardboard box containers. File users punch two holes in the sheets of paper and place them into rings connected to the lever arch file. These lever arch files are stored in an upright manner on shelves and are pulled off the shelves by the way of a small hole in the vertical backside of the folder; the vertical backside of the folder is also used for labeling (this labeling system has a larger surface area than the tab in American folders) the file as well. Furthermore, the labels in the lever arch files are always visible, while American file labels are obscured being stored in drawers or filing cabinets (Duncker, 2002).

Considering these differences, one could imagine how the desktop metaphor developed in the United States would not provide visual cues that are readily recognizable in other countries. If the filing metaphor, as it is perceived in the United States, was designed into a computer interface for a Japanese target user group, would they be able to draw the appropriate meaning from the metaphor? No. Cultural differences or meanings among groups of people are rooted in their history. Conflicts in metaphor meaning from group to group result from employing different cognitive, emotional, behavioral, and social structure processes. Therefore, it is essential to use metaphors that maintain relevance when the users consider the related physical object or source domain (Duncker, 2002). Other questions and answers to consider are: Is it likely that the target user group would experience major difficulty when trying to understand how to store their files? Yes.

While that desktop metaphor is a very simple example, there could be other instances when a software interface or website, intended for international markets, is developed with culturally invalid interface design elements. The United States has produced the majority of software for the international markets (Li et al., 2007). As a result, these interfaces are comprised primarily of American metaphors, representations, navigation schemes, and color uses. True globalization and localization of computer interface cannot be attained by simply translating idiomatic expressions and icons from one cultural situation to another (Duncker, 2002; Del Galdo, 1996). Cultural differences in metaphor comprehension must be equated to more than the superficial meanings of color and shapes of icons.

As we discuss the utility of marrying appropriate cultural factors to relevant interface design elements, we will recognize the duality of benefits to the end user and the sponsor of web-based and software applications. In any interface design venture there are at least two stakeholders: the design sponsor and the end user. There may certainly be more stakeholders associated with a design endeavor; however, the sponsors' ultimate goal is going to garner increased profitability and/or increased effectiveness of the services provided. Whether the sponsors are businesses or civil service providers, the success of website or software application is predicated upon usability, creditability, and trustworthiness of the application (Kondratova et al., 2005). As interfaces are disseminated globally, "culturability" is essential, i.e., understanding what comprises usability and creditability for a given application must be culturally motivated (Barber and Badre, 1998).

DESIGN FRAMEWORK FOR CULTURALLY INFORMED COMPUTER INTERFACES

An increasing number of empirical studies explore the relationships between cultural attributes of target user groups and the design and evaluation of computer interfaces. More researchers and practitioners are investigating the manner in which designers and developers can functionally translate cultural knowledge into interface design for websites, software packages, and mobile devices.

There are numerous interface and systems design methodologies utilized within the various software and computer interface design communities. Among the older software engineering techniques are the commonly known waterfall (Royce, 1970) and spiral (Boehm, 1988) methods. The waterfall method uses a top-down approach and a life cycle where the designers sequentially progress through a number of stages to include concept development, requirements specification, design, design review, implementation, integration and testing, and deployment. The spiral method takes into account large and complex systems where it is not feasible to sequentially progress through the top-down waterfall method. The spiral method emerged from the necessity for iteration in the software development process. In the spiral methodology, designers progress through the entire top-down process with multiple passes, extending the circle (the cycle nature of the process) to include more system details (Hartson and Pyla, 2012). Among the more recent approaches, as discussed in Hartson and Pyla (2012), is the agile software engineering method. The agile method focuses on delivering small software releases to customers after each short development cycle. The short turnaround releases are designed to provide functional pieces of software to the customer to use immediately. Over the course of successive iterations, the small but distinct pieces of software are implemented and become a more functional delivery of the whole system.

Unlike traditional waterfall-type methodologies, coding begins very early in the agile process. Moreover, the agile method uses a minimal amount of requirements specification and begins coding almost immediately. The overarching goal of the agile technique is to avoid the idea of heavy up-front design and get code written quickly. In addition, the agile software engineering method uses sprints as a mechanism for quick turnaround software releases. Developers rely on continuous communication with the customer in order to resolve any design problems encountered, as they develop the various features of the overall system in a piecemeal manner (Hartson and Pyla, 2012). Agile sprints comprise a series of steps: acceptance test creation, unit code test creation, implementing coding, code testing, and acceptance testing and deployment.

Hartson and Pyla (2012) identify several limitations of the agile method in regards to user experience. The agile approach in its purest form is code driven and does not incorporate any measures to ensure effective usability or an enhanced user experience. The user interface simply emerges as a result of the whims of a coder. Hartson and Pyla, recognizing the necessity for integrating the user experience into the design process, provide a synthesis of the usability and user experience methods with the agile methods. The specific details of that fusion are not discussed at length in this chapter. However, it should be noted that their integration of agile methods and user experience principles includes an abbreviated contextual inquiry and contextual analysis that may not offer the designers ample time to capture critical cultural factors for the interface design.

Other design approaches that have been used for interface development include case-based reasoning and use-case reasoning. Case-based reasoning is a tool that can be applied to various types of design applications (www.aiai.ed.ac.uk/links/cbr .html#intro). The key to case-based reasoning is the library of previous design cases (Maher et al., 1995). These cases are defined in terms of a problem description, a solution, and/or an outcome. The reasoning process is not included in these case descriptions but is intended to be inferred, based on the solution presented (www .aiai.ed.ac.uk/links/cbr.html#intro). In case-based reasoning solutions, old design cases are considered in order to determine feasible solutions for new design problems (Maher et al., 1995). Case-based reasoning functions only as a tool for retrieving previous cases; it would have to be supplemented with other types of reasoning in order to support the full design process.

Using case modeling (Bitter and Spence, 2003) as a methodology takes us a step toward a more user-centered focus. Use cases typically contain a bulleted list of steps necessary for an individual to achieve a very specific task with a given system. For example, the use case could describe the process flow for getting cash from an automated teller machine. The primary purpose is to help system developers form a conceptual model of a given system. Furthermore, the goal is to map the end users' and/or stakeholders' mental model with that of the system developer.

Each of the previously described design methodologies, with their distinct processes, characteristics, and proclivities toward end users, fall short of the up-front user focus necessary to design web-based and software applications, using globalization or localization paradigms. These design methodologies do not offer an opportunity for the designers to understand the user and their ecocultural setting with any real depth. Design for use across cultures will require a greater initial investment for the up-front user requirements.

As one considers the 21st-century opportunities to reach across continental borders and extend business, education, and recreation ventures into other countries and cultures, it is vital to approach design first from a user-centered versus a designer-centered perspective. Only then can designers, business owners, educators, and other design stakeholders effectively infuse their products into other cultural environments with any success and longevity. Furthermore, the interface and system design methodologies discussed previously do not engage users in the design process or seek to understand the larger context of use with much depth. The user's involvement in the use case process is limited to participating in sessions designed to verbally elicit system requirements, and there is no direct user involvement in the case-based reasoning approach.

Conversely, Rosson and Carroll's (2002) Scenario Based Development (SBD) user-centered interface design methodology presents opportunities for designers to acquire the more detailed aspects of the user, user's tasks, and their cultural and ecological settings. It also allows users to participate in the iterative refinement of the system's overall design. SBD is the foundation for the cross-cultural design model presented in this chapter. SBD was selected because of its (a) iterative user-centered design approach, (b) inclination toward the use of more naturalistic and/or ethnographic methods for understanding user preferences and requirements, and (c) use of scenarios. SBD's approach for knowledge gathering provide a basis for constructing interface designs that have been established upon real-world actions and meanings (Dourish, 2001).

SBD's framework and discussion offers a strong foundation for not only user-centered design steps, but lays the groundwork for eliciting culturally relevant information concerning the end users and integrating that information into culturally informed computer interface designs. Other user-centered design methodologies exist, but SBD provides an advantage in that it incorporates a broad spectrum of user considerations, which could have cultural implications, into the core components of its iterative design process. The categorization of its design phases offer a piecemeal, yet insightful way to parse out user-centered attributes that are essential to effective cross-cultural design—whether the focus be globalization or localization.

Furthermore, the remainder of this section highlights the Acculturalization Interface Design (AID) framework (Johnson, 2008), which extends Rosson and Carroll's (2002) SBD user-centered interface design concept toward more intently facilitating cross-cultural interface design. The extended and integrated interface design methodology presented here centers on merging (1) the SBD design phases and methodology, (2) various elicitation methods for determining user needs and requirements, and (3) noted design considerations for geographically and culturally dispersed people groups. The following discussion intends to achieve two primary goals. The first is to expand Rosson and Carroll's SBD to include discussion of methods for globalization and localization efforts, explaining the limitations of SBD and introducing culture-centered design interventions. The second goal is to utilize SBD's design model as a means of categorizing empirically based approaches and tools for culture-centered design to specific design phases.

The AID design framework is a step toward adapting the "culture" of the current interface design community into one that is more effective for the global community. This framework is titled the Acculturalization Interface Design because it advocates designers adopting techniques, methods, and tools that align with the cultural norms and expectations of target user groups. Its purpose is to supply members of the current interface design communities with the tools to extract essential cultural traits and social patterns of target people groups and develop them into culturally compliant interface designs.

The overarching goal of the AID framework is to be a repository of sorts that offers designers options for building in the necessary cultural factors into the entire user-centered design process. Thereby, the AID framework also serves to provide a systematic process for developing culturally appropriate interface designs, providing interface designers and design companies a framework that facilitates design for divergent user populations. The AID framework's added value is that it provides more details regarding how to collect the culturally relevant information and convert that information into culturally informed designs. It provides enhanced discussion about the conversion of cultural information into design, but maintains the flexibility needed to be a practical tool for a wide range of design endeavors.

Every interface design project is unique: each project has its own purpose, user group, stakeholders, goals, features, etc. So, it is most practical to provide designers with a design framework that presents the tools, methods and approaches in a la carte manner. The AID framework will elucidate how and when to incorporate cultural knowledge into the unique stages of the interface design process. The discussion is organized according to the design phases that the SBD uses: Analyze, Design (Activity Design, Information Design, and Interaction Design), and Prototype and Evaluate.

COMPUTER INTERFACE DESIGN PHASES

SBD discusses the use of various methods to include the user in the actual design phases. Table 4.1 provides an amalgamation of methods, means, and approaches for acquiring and organizing attributable and cultural information for the design's end-users. This collection of methods, means, and approaches is an aggregated list from SBD and a number of empirical studies exploring the cross-cultural computer interface design.

In the analysis phase, SBD guides the designer to begin understanding the users, their characteristics, and the context within which the users perform their work. The various methods listed include user activity analysis, artifact analysis, field studies, and ethnography, to mention a few. These methods/means, especially those

TABLE 4.1

A.I.D. Aggregation of Methods/Means/Approaches for Cross-Cultural Interface Design

SBD Design Phase	Method/Means/Approach
Analyze	User activity analysis
	User artifact analysis
	Social context analysis
	Field studies
	Observations
	Interviews
	Ethnographies
	Scenarios
	Foraging and local website audit (Badre, 2000; Smith et al., 2004)
	Cultural marker identification (Badre, 2000)
Activity design	Cooperative design
	Participatory design
	AID method (Johnson, 2008)
	Scenarios
Information design	User information perception
	User interpretation of information
	Cultural attractor identification (Smith et al., 2004)
	Making sense of the information
	Consistency and coherence
	Cultural marker pattern identification (Badre, 2000)
Interaction design	Norman's (1988) Gulf of Execution framework
Prototyping	Explore user requirements
	Storyboard technique
	PICTIVE method (Muller, 1993)
	Web crawler
	Usability testing
Usability evaluation	

Source: Adapted from Rosson, M. B., and Carroll, J. (2002). *Usability Engineering.* New York: Morgan Kaufmann.

elicitation methods that occur in the user's natural environment, are paths toward greater understanding and more generalizable characteristics of the user group(s). In the activity design phase, the user is given an opportunity to participate in the actual design of the computer interface. The user's active participation in the development process increases the likelihood that the user's mental model of the computer interface will match the system's design, thereby, providing the user with a product that matches his or her expectations. In information design and interaction design, the designer develops the information and interaction of the interface in such a way that fosters good user perception and interpretation. In prototype design, the designer uses methods that help define an effective low-to-moderate fidelity prototype for usability testing.

SCENARIOS

Go and Carroll (2004) identify a scenario as a description that contains (1) actors, (2) background information about the actors and assumptions about their environment, (3) goals or objectives of the actors, and (4) sequences of events and actions. Scenarios, of course, can be expressed in terms of a textual narrative; however, scenarios are produced as storyboards, video mock-ups, or scripted prototypes. Furthermore, these scenarios can be developed in formal, semiformal, or informal notation (Go and Carroll, 2004). Below is a simple example of a narrative (problem) scenario:

> Marissa was not satisfied with her class today on gravitation and planetary motion. She is not certain whether smaller planets always move faster, or how a larger or denser sun would alter the possibilities for solar systems. She stays after class to speak with Ms. Gould, but she isn't able to pose these questions clearly, so Ms. Gould suggests that she re-read the text and promises more discussion tomorrow. (Rosson and Carroll, 2002, p. 2)

Scenario use in human–computer interaction usually represents "a day in the life" of the user or potential user, as opposed to a larger scope like "a year in the life" (Go and Carroll, 2004).

In SBD, scenarios are developed and analyzed to assist designers with requirements for all phases of the design: analysis, system functionality, information presentation, interaction methods, documentation, the design of the prototypes, and the usability evaluations. SBD employs three major phases: analysis, design, and prototype and evaluation. Rosson and Carroll's (2002) SBD occurs in the order shown in Figure 4.1.

The analysis phase begins the process with requirements analysis and problem development. In this analysis, the problem and its context are studied using interviews with the users (including clients and stakeholders), field studies of the current practices, and brainstorming with both users and developers. The findings from this analysis become input for the scenarios (often in an iterative manner), which then demonstrate important characteristics about users, the tasks they perform, the tools they use, and the context within which they work. Employing scenarios in the cross-cultural design process offers designers a mechanism for further refining the design. At this point the scenarios are essential because they produce discussion about the current situation and how it works; they also raise questions about other situations and nuances that had not yet been considered. The discussion and results from the formation of scenarios are beneficial in that they help the developers to better envision the problem.

Rosson and Carroll's (2002) discussion about analyzing requirements suggest analyzing work practices and getting users involved in identifying their work-related needs. They recommend analyzing work practices in terms of (1) activities, (2) artifacts, and (3) social context. Considering these activities, the developers would ask

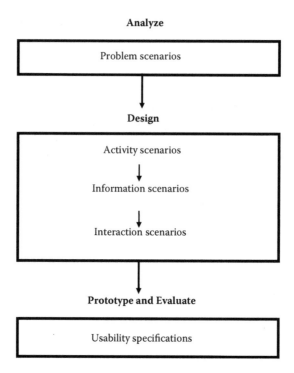

Analyze

Problem scenarios

Design

Activity scenarios

Information scenarios

Interaction scenarios

Prototype and Evaluate

Usability specifications

FIGURE 4.1   Scenario-based development. (Adapted from Rosson, M. B., and Carroll, J. (2002). *Usability Engineering*. New York: Morgan Kaufmann.)

questions concerning the organizational and/or personal goals that users pursue, and the actions they perform to achieve these goals.

One important addition to SBD's analysis phase is the acquisition of localized cultural information. For website application Badre (2000) highlights a three-stage process for creating cross-cultural websites: Foraging, Cultural Marker Identification, and Pattern Identification. Badre begins this process with the foraging stage. Foraging involves categorizing a large number (e.g., hundreds) of websites by country, genre, and language as the first step toward organizing significant cultural information for cross-cultural website design. The criterion for selection is that the website be in the country's native language. Badre surmises that websites written in the country's native language will be much more likely to incorporate culturally specific design elements—cultural markers. Cultural markers are defined as

> interface design elements and features that are prevalent, and possibility preferred, within a particular cultural group. Such markers signify a cultural affiliation. A cultural marker, such as a national symbol color, or spatial orientation, for example, denotes a conventionalized use of the feature in the web-site, not an anomalous feature that occurs infrequently. (Badre, 2000, p. 1)

These websites are scrutinized carefully for potential cultural markers and cross-listed by country and genre. Example genres include education, business, government, news and media, travel, health, science, etc. Cultural markers are identified based on their widespread occurrence for a specific cultural group and their minimal or non-existent occurrence in other groups.

Smith et al. (2004) propose an audit to conduct a very similar process of categorizing of local indigenous websites for cross-cultural design. They administer an audit to acquire cultural attractors, which is discussed later in this chapter, for interface design. They, too, select only websites that are written in the country's native language. These audits are generally conducted by a usability expert that is a member of the target group and has a deep cultural understanding of the group from personally experience or from personal ties.

ACTIVITY DESIGN

Second in the SBD is the design phase. The design stage addresses three substages: activity design, information design, and interaction design. Activity design is concerned with identifying the basic ideas and services of the new system. The final goal in this substage is to determine the system's functionality. Functionality specifications would include the types of operations that could be performed and the result of those operations.

SBD highlights the importance of designing effective, comprehensible, and satisfying activities. Activity design is the first substage within the SBD's design stage. The scenarios in activity design provide reasoning about which features of the activity are best suited for computer-based application. In these scenarios, the designers can explore the possibilities of new technology within the work related context. In addition, the scenario development and transformation will help provide ideas about the appropriate level of design generality. The discussion about designing comprehensible activities further supports the emphasis on specifying system functionality. The key to presenting comprehensible activities is to specify system models that match the user's mental models. The final consideration for the quest to determine system functionality is to design satisfying activities. It is important that users are able to perform activities on the computer that are of value (not just tedious, irritating, or uncommon tasks).

Participatory Design

Johnson (2008) uncovered another approach for identifying the types of activities (interface features and functionality) that users would fine meaningful. Johnson successfully administered an integrated participatory design methodology that was a fusion of a focus group, design ethnography, and participatory activity design, with a group of young African American novice computer users. She used this methodology as an approach for requirements analysis and activity design for users who are not typically represented during the interface design and development process. This particular study and project focused on designing a computer interface for economically underserved, computer-illiterate young adults. These young adults lived in an economically depressed region of a major city in the United States. Johnson's goal was to design a methodology that would be engaging, culturally relevant, and highly productive for generally marginalized populations in the United States. This methodology served as the mechanism for garnering culturally appropriate interface metaphors and interface design for a word-processing task.

SBD was amended here as well to address the needs of an economically under-served population of computer-illiterate individuals and to support the development of a computer interface that utilized innovative, yet culturally suitable computer interface metaphors. Her approach fused both the Analyze and Activity design SBD phases into one, with the use of participatory design and design ethnography. The focus group that she administered as a field study became the foundation for obtaining (a) task-based information about the users' activities and social context, (b) user preferences, (c) user expectations, and (d) a hand-drawn design of the users' desired interface. Johnson's (2008) augmented analysis and activity design procedures include the following addendums to SBD:

1. Specified the task(s) that would be implemented with computer applica-tion. Gave participants an opportunity to reflect on the task done by hand. Encouraged them to photograph (Johnson and Griffin, 1998; Rosson and Carroll, 2002) the key elements associated with completing that task by hand.
2. Conducted videotaped focus groups with participants; the ultimate goal of the focus group was to uncover the salient design (i.e., interface metaphors) features for the interface. These focus group recordings were used to con-struct design scenarios to aid interface design development. In addition, conversation about what they photographed was integrated into the focus group discussion to uncover any notable themes.
3. Performed a basic task analysis, artifact analysis, and theme analysis with the participants of the focus group. The emerging themes became the inter-face metaphors used in the design.
4. Instructed the users to draw how the interface should look based on the focus group discussions (Muller, 1993; Rosson and Carroll, 2002).

Johnson's participatory design methodology resulted in an interface design proto-type that would be later used for a usability study examining the influence of culture-specific design attributes on user performance. Each component of the interface's design was a direct result of her participants' cultural preferences for the design. Figures 4.2 and 4.3 illustrate the interface design with the interface metaphor that her participants mapped out for a word-processing task during the focus group. The fig-ures show a medium fidelity mock-up of the interface they designed in two different user-induced states. The overall metaphor is a "comfortable bedroom" that includes a radio, television, their bed, a table, and decorative wallpaper.

INFORMATION DESIGN

In the second substage of information design, the ultimate goal "is to support the per-ception, interpretation, and comprehension of computer-based information" (Rosson and Carroll, 2002, p. 115). Perceiving information is concerned with creating a design where viewers can clearly distinguish structures in an information display. In an information display, the structures are assembled with pixels, colors, or tones.

FIGURE 4.2 *African-American-inspired interface: Comfortable Bedroom Metaphor.* (From Johnson, K. [2008]. Process, Preference, and Performance: Considering Ethnicity and Socio-Economic Status in Computer Interface Metaphor Design. Unpublished doctoral dissertation, Virginia Tech, Blacksburg.)

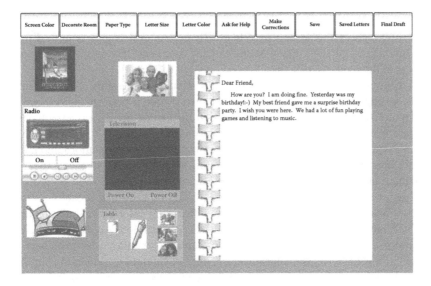

FIGURE 4.3 *African-American-inspired interface: Comfortable Bedroom II Metaphor.* (From Johnson, K. [2008]. Process, Preference, and Performance: Considering Ethnicity and Socio-Economic Status in Computer Interface Metaphor Design. Unpublished doctoral dissertation, Virginia Tech, Blacksburg.)

Moreover, SBD put forward several principles to follow for constructing perceivable information structures. First, it highlights Gestalt principles. These principles describe the architectural properties of visual information. More specifically, they describe "how individual bits of information are grouped together, what elements will be seen as a coherent figure, and what elements will appear as background" (Rosson and Carroll, 2002, p. 113). This principle emphasizes guiding rules about proximity, similarity, closure, area, symmetry, and continuity; all of which have an impact on an individual's low-level perception. Furthermore, Rosson and Carroll (2002) recommend using Gestalt principles for perceptual matters that occur at a higher level. High-level perception transcends the basic perception of pixels, colors, and shapes and is extended to the perception of menu bars, scroll bars, paragraphs, file lists, radio buttons, and icons.

Application of the Gestalt principles should be done with caution. For example, Cha and Nam (1985), as well as Choi and Nisbett (1998), found differences in the way westerners and East Asians make causal attributions and predications. Westerners and East Asians differ in that one group ascribes to categorization based on rules versus family resemblance, and categorization based on taxonomic labels versus relationships. Chiu (1972) adds that westerners group objects and events in a taxonomic manner, while East Asians' grouping of objects and events is relational. East Asians are adept at identifying relationships among objects in an environment and are less effective in distinguishing an object from its surroundings. Conversely, westerners tend to focus on salient objects embedded within an environment. Consequently, East Asians perceive more changes within the background or environment and westerners are more likely to detect changes with the central figures (Li et al., 2007). A culturally sensitive recommendation is to investigate and understand the implications of variances in broad psychological views of the intended user groups. Designers should be aware that different cultural groups require different perception principles for information design. These considerations inform essential aspects of the interface design.

Perception is one phase of human information processing. The next phase would be cognition, where reasoning and interpretation is performed (Wickens and Hollands, 2000). Perceiving the information display as objects or sets of objects allows for interpretation to take place (Rosson and Carroll, 2002). Therefore, the next portion of information design in SBD concerns "interpreting information." SBD defines interpretation as (the user) determining what the display elements mean and how these elements fit into the context of the interface. Rosson and Carroll (2002) elaborate on the concept of using elements of familiarity and meaning, such as words, images, and symbols to assist users with display information interpretation.

The "interpreting information" component of information design takes on greater cultural relevance with the identification of cultural attractors, which has been discussed in regards to cross-cultural website design. Cultural attractors are similar to cultural markers; however, the end product attaches a specific meaning to the identified signs or cultural symbols. Smith et al. (2004) define cultural attractors as "the interface design elements of the website that reflect the sign and their meanings to match the expectations of the local culture" (p. 70). An audit of prominent and successful (indigenous) local websites, as discussed previously, serves as the primary

ICICI Bank	Grand Commercial Bank
www.icicibank.com	www.grandbank.com.tw

FIGURE 4.4 Captured results from website audits. (From Smith, A. et al., 2004, *Interacting with Computers,* 16, p. 71.)

source for outlining the cultural attractors for a specified region. These cultural attractors are identified by applying a semiotic perspective to the examination of computer-based signs. Semiotics is a discipline that uses an understanding of acts of signification to connect meaning, mean-making, communicationing, and culture (as stated in Smith et al., 2004). In this particular context, computer-based signs include images, textual cues, icons, and sounds. It is proposed that both the signs and their meanings will differ from culture to culture.

Developing an in-depth understanding of how to create websites that equally and adequately address the users as well as the domain area (e.g., e-commerce, e-learning) is a result of thorough examinations of various signs in the local culture, the context of use, and the meanings the locals assign to them. Completed investigations generally yield the following cultural attractors: colors, color combinations, banner ads, trust signs, language cues, currency formats, metaphor use, navigation controls, and other visual elements that make up the basic "look and feel" of a website. Figure 4.4 provides snapshots of banking websites from India and Taiwan that were audited. Table 4.2 provides the summary of the cultural audit that was conducted.

A number of researchers have studied cognitive styles from the viewpoint that certain cognitive performances/abilities may be more or less critical, in particular, ecological and cultural contexts. Using an ecological context refers to considering the setting in which humans interact with their environment, and the set of resulting relationships that present a variety of life possibilities for a population. An ecological analysis of the demands of an environment would include investigating questions such as what are the necessary things that need to get done (in the environment) in order to survive and what cultural norms/practices or social situations lead to the development of the necessary cognitive performances (as stated in Berry et al., 1992). This eco-cultural framework consists of a system of variables segmented into two main classes: ecological context and sociopolitical context. Variables include psychological characteristics, including traits, motives, abilities, and attitudes, which result from the two transformative sets of variables that signify the influence or transmission from population variables to individuals. This conceptual model was designed to present the general relationships between classes of variables that can be used to explain the cross-cultural similarities and differences in human behavior.

TABLE 4.2

Cultural Attractors

Indian e-Finance Website Attractors

Website	Attractors	Rationale
ICICI Bank	Color red	Associated with vitality, energy, prosperity, health, ambition, and initiative
	Color saffron	Considered auspicious by Hindus, Sikhs, Jains, and Buddhists
	In combination	Signifies prosperity for current and future customers
	Use of language	
	Term *munshi*	In Hindi means "family accountant"
	Term *padres*	In Hindi means "abroad"
	Link *munshi–padres*	Combination signifies customer friendly
	Religious iconography	To seek blessings from God
	Charity puja service—enabling donations for good causes in India	Hindi communication with his or her personal Hindu God or Goddess
State Bank of India	Gandhi's image	Conjures up patriotic feelings
		Signifies that the bank serves all Indians whether at home or abroad
	Map of India	Bank serves all Indians, rural and urban

Source: Smith, A. et al., 2004, *Interacting with Computers,* 16, 72–73.

Other researchers have studied cross-cultural psychology from the basis of the eco-cultural conceptual framework. Studies that date back to the 1960s and 1970s, including the previously discussed works of Berry (1976) and Dawson (1967), are among those putting forth that ecology and culture are a source for psychological differentiation between cultural groups. Berry et al. (1976) studied two different groups in Africa. They compared the cognitive style of the African Pygmy (Blaka) hunters and gatherers with a sample of individuals from an agricultural village within the same geographic region as the Pygmy. They discovered that there was a difference in cognitive style (assessed by way of an African-embedded figures test) in the two African cultures, but only when differences in acculturation were considered as well.

John Ogbu also applies a cultural–ecological model in his cross-cultural psychology research (Ogbu, 1995). His work has been concentrated in the area of minority education from a cross-cultural perspective. Nevertheless, he, too, ascribes differences in intelligences (or cognitive performances) to the social or cultural adaptations characteristic to a given population. Ogbu asserts that children belonging to a given population are socialized to attain the cognitive skills that are essential for existing in that culture (Ogbu, 1994). Furthermore, Ogbu (2002) identifies a number of cultural amplifiers of intelligence, which he describes as those activities in an eco-cultural niche that necessitate enhanced intellectual skills attainment. An eco-cultural niche makes reference to the broad societal environment where cultural activities/tasks create cognitive problems for members of the population. For example, Ogbu identifies superior mathematical skills in children who grow up in cultures whose economy is rooted in commerce as opposed to farming. In this case, the cultural amplifier is commerce, and the skill it augments is mathematics (as stated in Obgu, 2002). Overall, this discussion is provided to offer a general framework for how culture can potentially be a mitigating factor for the development and execution of cognitive skills for a given ethnic/cultural group. Therefore, additional considerations for deep cultural understanding and interface design are cognitive style and the interaction between cognitive style and acculturation.

The last concern within SBD's information design phase is being able to make sense of the information that has been perceived and interpreted (Rosson and Carroll, 2002). Often, users will try to make sense of information by making associations with what they already know about their task and determining how it fits into the larger scheme of their desired goals or interests. When that does not work, users will consider alternate interpretations of the information and try to make sense of it that way. SBD suggests consistency in the application of colors, vocabulary, shapes, layout, visual metaphor, information models, and dynamic displays to facilitate making sense of the information. To extend SBD's guiding principle toward cultural relevance, designers must first identify the common uses and expectations of colors, shapes, metaphors, text layout, etc., and implement those with consistency across their designs.

While there are some obvious cultural issues that designers have to consider (e.g., differences in spoken language or dialect), there are some not as obvious cultural issues as well. For instance, in some countries the social norm is for both men and women to be tender, modest, and concerned with the quality of life (Hoftstede, 1997). So for them, it would be less appropriate to use terminology such as ABORT or KILL (Shneiderman, 1992) to describe a function to exit or end a program. This example is applicable for images and symbols as well.

Here again is an occasion for cultural awareness in design. According to Feurzeig (1997), an economically and educationally underserved group of inner-city high school students in Boston, Massachusetts, were capable of learning concepts related to complex Galilean relativity problems and genetics using educational computer programs just as well as high-school students from a more affluent background. However, the inner-city group had difficulty transferring their knowledge of those concepts to written tests. It was determined that the underserved students lacked the literacy and communication skills necessary to reproduce the complex concepts they learned into written material. Therefore, meager reading comprehension skills may hinder user performance with text-centered interface design. Designers must be conscious of the possibility that their target user group may not possess the literacy levels necessary to make sense of the presented information. So designers should design with simple and descriptive labels, titles, and phrases, as well as providing instructions when appropriate.

Another opportunity for cultural awareness for "making sense of the information" concerns visual metaphor. Metaphors are only valuable when they exploit the users' current knowledge about a task, process, or social norm. Cultural differences unaccounted for in the design may result in implementation of inappropriate metaphors. For example, Aaron Marcus suggests that cultures high in uncertainty avoidance would be most comfortable or reassured with an interface that uses simple and easily discernible metaphors (Marcus and Gould, 2001). Uncertainty avoidance is one of five cultural dimensions that Geert Hofstede uncovered during an anthropological study of a global business organization. Uncertainty avoidance is defined as the degree to which individuals feel threatened in the midst of unknown or uncertain situations (Hofstede, 1997).

Badre's (2000) cultural marker pattern recognition seeks to determine what level of interplay emerges from a cross examination of cultural markers, country, regions and genres. Identifying the interaction among these three variables provide deeper insight into how a designer should structure the information for more effective meaning-making for various countries and domain areas such as government, education, travel, health, business, and beyond.

Culture and Interface Design Elements

Computer interfaces generally comprise a combination of four design elements: metaphor, navigation, icons/symbols, and colors. These are the key components that establish the look and feel of interfaces. These key elements have appeared throughout this chapter's discussion. All cultural knowledge acquired concerning any specified user groups must be translated appropriately when using these design elements. While each of these elements are essential to the makeup of a design, interface metaphors, which are mechanisms for meaning-making, are often embedded within the other design elements. Subsequently, the integration of metaphors into design is potentially quite impactful.

Interface metaphors can be found in all types of computing systems and applications. For example, verbal and pictorial metaphors are pervasive in computing systems such as mobile phones, digital cameras, audio listening/recording devices, gaming devices, and beyond. Contemporary examples of pictorial and verbal metaphors embedded within a computer system are smart phones such as Apple's iPhone[TM] and the Droid by Motorola. In these examples the user's interaction is driven by pictorial metaphors (Johnson and Smith-Jackson, 2009). These types of phones possess a myriad of features and applications. Their main menu structures are formed primarily by pictorial metaphors that are used to depict functions, including email, messaging, voicemail, navigation, calculator, calendar, clock settings, the web, etc. Johnson and Smith-Jackson (2009) affirm that, "Considering what we know about culture's affect on cognition, transferring metaphor (pictorial) into other cultural domains is a questionable activity. The images, and even the words, represented can potentially take on variant meanings, thereby invoking erroneous user actions" (p. 624). The use of metaphors, images, and icons/symbols that lack culturally informed application may result in ambiguity and even offense among cultural groups, establishing sources for poor and dissatisfying user experiences.

INTERACTION DESIGN

The last substage in the design phase of SBD is "interaction design." Interaction design refers to the specification of mechanisms for gaining access to and manipulating task information (e.g., windows, icons, and menus). The ultimate goal of interaction design is to minimize the users' cognitive effort. In the discussion of interaction design, several antidotes for reducing cognitive effort are presented. One concern is to make action obvious to the user. In some cases direct manipulation (as in interaction style) can facilitate user action planning. In direct manipulation, the interface is constructed with objects and actions that map to real-world objects and actions. For example, the user can grab, drag, or stack a folder on the interface (Rosson and Carroll, 2002). The key here is to make interface interaction discernable by implementing analogies to the real world. Cultural awareness would dictate that designers develop "real world" semantic and physical analogies that are indicative of the culture that has shaped particular users. Gaps in analogy comprehension may produce user performance problems.

Another component of this interaction design substage concerns the execution of the action. This stage focuses on selection of input devices for a task. Input devices include a mouse, keyboard, button, joystick, trackball, data glove, and finger/stylus for touch screens. The discourse on execution of the action concerns selecting input devices and optimizing performance with them. One brief point about interaction through input devices is that the designer will have to be certain that the input devices support functions that are consistent with their cultural practices. For instance, Katre (2006) completed a cross-cultural usability study of bilingual (Hindi and English) mobile phones. The study focused on keypad design, positioning of Devanāgari alphabets, font types, text entry techniques among other things. Katre determined that the keypad layout should remain consistent with the original structure of language in its purest form, without any deviations.

PROTOTYPING

The final stage of the overall SBD is prototyping and evaluation. Prototypes are developed and tested to support the iterative portion of the SBD process. Prototypes are used to determine and refine user needs, explore design possibilities, perform participatory design, and/or explore open issues.

Prototyping Tools for Cross-Cultural Design

Kondratova and Goldfarb (2005) have conceptualized a cross-cultural prototyping tool that will serve as a functional repository of cultural/country data for building culturally appropriate interface design alternatives. They envision this tool being a mechanism for developing a first draft of a culturally informed interface design for a selected country. The process for developing a first draft design from the repository will begin with the "look and feel" advisor. The "look and feel" advisor would step the designer through a process in order to determine the basic "look and feel" of the interface. The steps would comprise of selecting (1) a geographical region, (2) a country, (3) a predefined color and style selection for the country, and (4) predefined layouts, graphics, typography, etc. The "look and feel" advisor works in tandem with the building blocks repository, which consists of Visual Cultural User Interface Templates, visual design objects, and cultural data objects. The "building blocks" repository serves as supplementary information, used to further refine the design. Cultural data objects refer to things such as currency, date/time formats, and audio objects.

Both the Visual Cultural User Interface Templates and the cultural building blocks repository are both populated by completing a "cultural audit" via a Web crawler. The Web crawler is designed to extract the predetermined cultural information, which is in turn assigned appropriately as cultural markers. The building blocks also serve as reusable libraries for instigating the design of localized of websites.

This cross-cultural prototyping tool has been introduced here more as a concept than a recommended resource. The literature and discussion in this chapter brings to light many of the challenges related to cross-cultural interface design. There are so many variables in these types of designs that it is not likely that a computer-based tool alone will capture the cultural nuances that designers will need for effective design. Here, the human element is essential for synthesizing relevant findings and vetting all cultural design considerations through its very specific context of use. However, the central concept of developing a cross-cultural design repository is one that needs to be adopted by researchers and practitioners.

USABILITY EVALUATION

Cultural differences have implications for both the design and the process for design. We cannot make the assumption that Western strategies, methodologies, and techniques for user-centered design and participation are applicable in other cultures or within multicultural teams without any adaptations (Smith et al. 2004).

Simply stated, "usability measures the quality of a user's experience when interacting with a product or system—whether a website, a software application, a mobile technology, or any user-operated device" (http://usability.gov/basics/index.html). Usability.gov, much like many other organizations, researchers, and practitioners that participate in computer interface design, think of usability as a combination of five key factors: ease of learning, efficiency of use, learnability, retainability, and user satisfaction (Hartson and Pyla, 2012). While these are widely accepted and referenced attributes of usability, researchers and practitioners often apply a Western-centric approach to the evaluation of these usability factors. The notions of efficiency, effectiveness, and even user satisfaction are very different across cultures. Therefore, usability must be measured in culturally relevant ways.

Usability evaluation is any analysis or empirical study of a prototype's ease-of-use (Rosson and Carroll, 2002) and usefulness (Hartson and Pyla, 2012). Usability testing can be performed for formative or summative evaluation. At this stage in the process, it may seem that the culturally based interventions would be unnecessary. Notably, empirical research shows that culture plays a significant role in usability evaluation process and methods in addition to the design stages of the user-centered design process. These studies show that culture affects the think-a-loud protocol (Yeo, 1998), structured interviews (Vatrapu and Perez-Quinones, 2006), questionnaires (Day and Evers, 1999), and focus groups (Beu et al., 2000).

In a relevant study, Vatrapu and Perez-Quinones (2006) report that Indian participants in a usability study using structured interviews found more usability problems and made more suggestions to an Indian interviewer than to a foreign

(Anglo-American) interviewer. They provide several key takeaways from their study: (1) cultural mismatches between the interviewer and the user affect outcomes of structured interviews, (2) interviewers from the same culture may be more effective in obtaining usability problems in high-power-distance cultures, and (3) Hofstede's cultural dimensions model can facilitate the selection of usability evaluation methods for cross-cultural usability testing. Furthermore, as the business ventures and other entities continue to broaden their reach into more culturally and geographically dispersed locations, understanding the role of culture in the implementation of remote usability comes into play (Hartson, Castillo, Kelso, Kamler, and Neale, 1996).

CASE STUDY: USABILITY PROBLEMS WITH INTERFACE METAPHOR MISMATCHES

Metaphor mismatches are the nemesis to good interface design and can be the very agent that disenfranchises users and potential users of various cultural groups. Duncker (2002) performed a cross-cultural usability study of the "library" interface metaphor with white New Zealanders of European descent and the Maori people who are indigenous to New Zealand. To understand the cultural schemas, the researcher studied different aspects of Maori culture. The study of the Maori uncovered information about their history, values, customs (including the way they pass on knowledge from generation to generation), and perceptions about the westernized library system that is used in New Zealand. When the Maori, who have distinct perspectives about the purpose of/service in a physical library, were asked to perform a number of tasks using a library interface metaphor, various issues arose. Those issues later became usability problems for the Maori students when executing tasks using the digital library designed according to westernized standards.

Among Duncker's (2002) list of general Maori character traits (where some are general and others are specific to digital libraries) is their inclination for

- Being oriented toward collectivism and tribal unity
- Holding their genealogy, sacred objects (of their culture), tribal privacy, and property rights in great esteem
- Being oriented toward the past (They view the past as the forward direction and pay no attention to the future. They see time progressing toward the past and those of a westernized culture view time as progressing toward the future.)
- Believing that representations of people (whether in text, pictures, or carvings) are very sacred and should only be used in their sacred tribal environments
- Being partial to face-to-face communication
- Disagreeing with the openness of the Web (concerning Maori content or information)
- Feeling unwelcome in the library, even when the library staff is friendly by Western standards
- Disagreeing with the use of the English classification systems for Maori content
- Being unfamiliar with publication formats: journals, series, and proceedings

The Duncker (2002) study was performed in New Zealand using a local college's digital library. Using the digital library (and western libraries in general) proved to be a difficult undertaking for Maori college students. The cultural characteristics described above were in most cases the root of the Maori's usability problems with the digital library. Another potentially relevant issue for the Maori students is that the libraries in New Zealand, which are westernized, are often a repository for Maori artifacts—artifacts that include Maori genealogies and other aspects of their history. The Maori sometimes feel that the history of their people does not receive the respect that it deserves and should not be available for viewing by the general public.

The specific usability problems that surfaced from Duncker's study were understood to have the following foundations:

1. Libraries emphasize individualism, and the Maori culture is characterized as being collectivist in nature.
2. Maori are partial to face-to-face communication, which is not generally supported in western libraries.
3. Maori value their sacred objects, tribal privacy, and property rights; library policies generally do not reflect those values.
4. Maori do not agree with the openness of the web.
5. Maori do not feel welcome in the library, even when the library staff is friendly by western standards.
6. English classification systems are not appropriate for Maori historical content (they are not familiar with the western/Anglo-American formats and categorization).
7. Maori have trouble with publications formats: journals, series, proceedings (they do not generally know what these are).

Duncker (2002) found that the Maori are able to work with digital libraries, but certain aspects of the library metaphor break down. These breakdowns made working with digital libraries arduous for the Maori and fraught with negative critical incidents. Below is a negative critical incident occurring during a usability session with a Maori student that demonstrates usability problem 7.

Participant stops working. Silence.
Researcher: "What is the matter?"
Participant: " I don't know."
Researcher: "What are you doing?"
Participant: "I am not sure. Maybe go there?" (Points at the link Journal and Proceedings)
Participant: "Perhaps?" (Looks at the researcher)
Researcher: "It depends on what you want to do."
Participant is silent and looks at the screen.
Researcher: "Do you know what journals and proceedings are?"
Participant: "No."
Researcher explains what journals and proceedings are.
Participant carries on. (Duncker, 2002, p. 228)

CONCLUSION

As the various design communities continue to reach out internationally, they must remain user-centered in approach and apply any relevant design methodologies and tools in the most sensible manner for their design projects. It is not likely that any one cultured-based design tool alone will garner the essential factors required for all types of design endeavors. The purpose of this chapter was to provide a high-level user-centered design framework that incorporates practical tools and techniques to facilitate cultural relevance in design for globalization and localization endeavors.

The Acculturalization Interface Design (AID) methodology seeks to achieve that purpose as it merges (1) the Rosson and Carroll's (2002) Scenario Based Development design phases, (2) various elicitation methods for determining user needs and requirements, and (3) noted design considerations for geographically and culturally dispersed people groups. Furthermore, the AID methodology is an amalgamation of methods and tools that serve as an initial point of interface design conceptualization and development. The AID methodology also serves as a framework for categorizing empirically based approaches and tools for culture-centered design to specific phases of the interface design process. Practitioners, designers, and researchers should assess the impact of culture on their design projects individually, in order to integrate the appropriate tools and methods necessary to capture any unique cultural nuances for a specific user group. In doing so, design communities will be able to maximize their users' experience, and continue to grow a repository of culturally based design solutions for others to reference in the future.

REFERENCES

Badre, A. (2000). The Effects of Cross Cultural Interface Design Orientation on World Wide Web User Performance. GVU Tech Reports, Retrieved from http://www.cc.gatech.edu/gvu/reports/2001.

Barber, W. and Badre, A. (1998). Culturability: The merging of culture and usability. Presented at the Conference on Human Factors and the Web, Basking Ridge, New Jersey: AT&T Labs. Retrieved from http://zing.ncsl.nist.gov/hfweb/att4/proceedings/barber/.

Berry, J. W. (1976). *Human Ecology and Cognitive Style*. New York: Wiley and Sons.

Berry, J. W., Poortinga, Y. H., Segall, M. H., and Dasen, P. R. (1992). *Cross-cultural psychology: Research and applications*. New York: Cambridge University Press.

Beu, A., Honold, P., and Yuan, X. (2000). How to Build Up an Infrastructure for Intercultural Usability Engineering. *The International Journal of Human-Computer Interaction, 12* (3&4), 347–358.

Bitter, K. and Spence, I. (2003). *Use Case Modeling*. Boston: Addison-Wesley.

Blomberg, J., Burrell, M., and Guest, G. (2003). An Ethnographic Approach To Design. In J. A. Jacko, A., and A. Sears (Ed.), *The Human-Computer Interaction Handbook* (pp. 964–990). Mahwah, NJ: Lawrence Erlbaum Associates.

Blomberg, J., Giacomi, J., Mosher, A., and Swenton-Wall, P. (1993). Ethnographic Field Methods and Their Relation to Design. In D. Schuler and A. Namioka (Eds.), *Participatory Design: Principles and Practices*. Hillsdale, New Jersey: Lawrence Erlbaum Associates.

Boehm, B. W. (1988). A spiral model of software development and enhancement. *IEEE Computer,* 21(5), 61–72.

Cha, J. and Nam, K. (1985). A test of Kelley's cube theory of attribution: A cross-cultural replication of McArthur's study. *Korean Social Science Journal,* 12, 151–180.

Choi, I. and Nisbett, R. (1998). Situational salience and cultural differences in the correspondence bias and in the actor-observer bias. *Personality and Social Psychology Bulletin,* 24, 949–960.

Chiu, L. (1972). A cross-cultural comparison of cognitive styles in Chinese and American children. *International Journal of Psychology,* 7, 235–242.

Dawson, J. L. M. (1967). Cultural and physiological influences upon spatial-perceptual processes in West Africa. *International Journal of Psychology,* Part 1(2), 115–128.

Day, V. and Evers, V. (1999). Questionnaire Development for Multicultural Data Collection. In E. del Galdo and G. Prahbu (eds.), *Proceedings of the International Workshop on Internationalisation of Products and Systems,* Rochester, NY.

Del Galdo, E. M. (1996). Culture and Design, In Elisa M. Del Galdo and Jakob Neilsen (eds.), *International User Interfaces.* New York: John Wiley & Sons.

Dourish, P. (2001). *Where the Action Is: The Foundation of Embodied Interaction.* Cambridge, Mass: The MIT Press.

Duncker, E. (2002). *Cross-Cultural Usability of the Library Metaphor.* Paper presented at the JCDL '02, Portland, Oregon.

Feurzeig, W. (1997). Extending knowledge access to underserved citizens. In N. R. Council (Ed.), *More than screen deep: Toward every citizen interfaces to the nation's information infrastructure* (pp. 395–402). Washington, D.C.: National Academy Press.

Go, K. and Carroll, J. (2004). The Blind Men and the Elephant: Views of Scenario-Based System Design. *Interactions,* 11, 45–53.

Gomez, L. M., Egan, D. E., Wheeler, E. A., Sharma, D. K., and Gruchacz, A. M. (1983). *How interface design determines who has difficulty learning to use a text editor.* Paper presented at the Proceedings of the CHI '82 Conference on Human Factors in Computing Systems.

Hartson, H. R., Castillo, J. C., Kelso, J., Kamler, J., and Neale, W. C. (1996). Remote Evaluation: The Network as an Extension of the Usability Laboratory. *Proceedings of CHI '96 Human Factors in Computing Systems,* 228–235.

Hartson, R. and Pyla, P. (2012). *The UX Book: Process and Guidelines for Ensuring a Quality User Experience.* Waltham, Boston, Elsevier, Inc.

Hofstede, G. (1997). *Culture and Organizations: Software of the Mind.* New York: McGraw Hill.

Johnson, K. (2008). *Process, Preference, and Performance: Considering Ethnicity and Socio-Economic Status in Computer Interface Metaphor Design.* Unpublished doctoraldissertation, Virginia Tech, Blacksburg.

Johnson, K. T. and Smith-Jackson, T. L. (2009). A human factors view of the digital divide. In E. Ferro, Y. Dwivedi, J. Gil-Garcia, and M. Williams (Eds). *Handbook of Research on Overcoming Digital Divides Constructing an Equitable and Competitive Information Society.* Hershey, PA: IGI Global, pp. 606–629.

Johnson, J.C. and Griffin, D.C. (1998). Visual data: Collection, analysis, and representation. In V. DeMunck and E. Sobo (Eds.), *Using methods in the field: A practical introduction and casebook* (pp. 211-228). Walnut Creek, CA: Altamira.

Katre, D. (2006). A Position Paper On Cross-Cultural Usability Issues of Bilingual (Hindi and English) Mobile Phones. *Indo-Danish HCI Research Symposium.*

Kondratova, I. and Goldfard, I. (2005). Cultural Visual Interface Design. *Proceedings of EdMedia 2005, World Conference on Educational Multimedia, Hypermedia and Telecomunications.* Montreal, Quebec, Canada. June 27–July2, 2005. pp. 1255–1262. NRC 48237.

Kondratova, I., Goldfard, I., Gervais, R., and Fournier, L. (2005). Culturally Appropriate Web Interface Design: Web Crawler Study. *Proceedings of the 8th IASTED International Conference on Computer and Advanced Technology in Education* (CATE 2005). Oranjestad, Aruba. August 29–31, 2005. pp. 359–364, NRC 48253.

Li, H., Sun, X., and Zhang, K. (2007). Culture-Centered Design: Cultural Factors in Interface Usability and Usability Tests. Eighth ACIS International Conference on Software Engineering, Artificial Intelligence, Networking, and Parallel/Distributed Computing. IEEE Computer Society, pp. 1084–1088.

Maher, M., Balachandran, M. B., and Zhang, D. M. (1995). *Case-Based Reasoning Design.* Mahwah, New Jersey: Lawrence Erlbaum Associates, Publishers.

Marcus, A. and Gould, E. (2001). Cultural Dimensions and Global Web Design: What? So What? Now What?, AM+A Publications, http://www.amanda.com/publications/.

Moggride, B. (1993). Design by story-telling. *Applied Ergonomics,* 24(1), 15–18.

Muller, M. (Ed.). (1993). *PICTIVE: Democratizing the Dynamics of the Design Session.* Hillsdale, New Jersey: Lawrence Erlbaum Associates.

Norman, D. A. (1988). The Design of Everyday Things. New York: Doubleday.

Ogbu, J. (Ed.). (1994). *From Cultural Differences to Differences in Cultural Frame of Reference.* Mahwah, New Jersey: Lawrence Erlbaum Associates.

Ogbu, J. (Ed.). (1995). *Origins of Human Competence: A Cultural-Ecological Perspective.* New York: New York University Press.

Ogbu, J. (Ed.). (2002). Cultural Ampliers of Intelligence: IQ and Minority Status in Cross-Cultural Perspective. Mahwah, New Jersey, Lawrence Erlbaum Associates.

Rosson, M. B. and Carroll, J. (2002). *Usability Engineering.* New York, Morgan Kaufmann.

Royce, W. W. (1970, August 25–28). Managing the development of large scale software systems. In Proceedings of IEEE Western Electronic Show and Convention (WESCON) Technical Papers (pp. A/11–9). Los Angeles, CA. (Reprinted in Proceedings of the Ninth International Conference on Software Engineering, Pittsburgh, ACM Press, 1989, pp. 328–338.)

Shneiderman, B. (1992). Designing the user interface: strategies for effective human-computer interaction (2nd ed.). Reading, MA, Addison-Wesley.

Smith, A., Dunckley, Lynne, French, T., Minocha, S., and Chang, Y. (2004). A process model for developing usable cross-cultural websites. *Interacting with Computers,* 16, 63–91.

Vatrapu, R. and Perez-Quinones, M. (2006). Culture and usability evaluation: The effects of culture in structured interviews. *Journal of Usability Studies,* 1(4), 156–170.

Veroff, J. and Goldberger, N. (1995). What's in a name. In N. R. Goldberger and J. B.Veroff (Eds). *The Culture and Psychology Reader.* New York, New York University Press, pp. 3–21.

Wickens, C. D. and Hollands, J. G. (2000). *Engineering Psychology and Human Performance* (3rd ed.). Upper Saddle River, NJ, Prentice-Hall.

Yeo, A. (1998). *Cultural Effects in Usability Assessment.* CHI 98:18–23.

5 Creating Inclusive Warnings

The Role of Culture in the Design and Evaluation of Risk Communications

Christopher B. Mayhorn,
Michael S. Wogalter, Richard C. Goldsworthy,
and Brannan R. McDougal

CONTENTS

CREATING INCLUSIVE WARNINGS: ROLE OF CULTURE IN THE DESIGN AND EVALUATION OF RISK COMMUNICATIONS

Warnings are risk communications used to inform people about hazards and to pro-vide instructions so as to avoid or minimize undesirable consequences such as death, injury, or property damage. Warnings are used in a variety of contexts for numerous kinds of potential hazards. For instance, a product warning might be used to inform users about the electrocution hazard associated with a kitchen appliance, whereas an environmental warning might be used to advise people to evacuate the area where a hurricane is expected to make landfall. While these examples of warnings might appear to be very different, they share a number of commonalities because they are both persuasive safety communications used to guide the behavior of those who receive them.

Based on the classic work of Lasswell (1948) and Hovland, Janis, and Kelley (1953), all persuasive communications should be analyzed in terms of *source* (the entity that initiates communication), *message* (content of communication), *channel* (how the message is communicated), *receiver* (target of the communication), and *effect* (desired behavioral change). These components of risk communications have been studied in depth over the past several decades (see Lindell and Perry, 2004; Wogalter, 2006 for extensive reviews). The present chapter focuses on one of these components, receivers. The characteristics of the person being warned are subdi-vided into topics that are discussed.

Although it is often recognized that warning effectiveness depends on the extent to which these risk communications have been designed to match the needs and capabilities of the target audience, it is equally important to understand that the characteristics of message recipients vary from one individual receiver to the next; therefore, warning designers need to understand that their target audience may not be homogeneous (Smith-Jackson, 2006a). For instance, a number of research-ers such as Goldhaber and deTurck (1988) and Flynn, Slovic, Mertz, and Carlisle (1999) have investigated the role of gender on warning compliance and risk percep-tion. Others have investigated chronological age as an individual difference when people encounter warnings and other risk communications (Mayhorn and Podany, 2006; Rousseau, Lamson, and Rogers, 1998; Young, Laughery, Wogalter, and Lovvoll, 1999). Unfortunately, not all receiver characteristics have been as exten-sively studied. In particular, there is a demonstrated paucity of research in the area of understanding how cultural attributes of receivers impact warning effectiveness (Reid, 1995; Smith-Jackson, 2006b). As will be discussed later in the chapter, the communication–human information processing model (C-HIP) will be used to expose the need for consideration of cultural ergonomics because there are serious gaps in the current warning literature.

CULTURE, SUBCULTURE, AND ETHNICITY: DEFINITIONS AND DISTINCTIONS

Perhaps one explanation for the relative lack of research regarding culture in this context comes from an incomplete (and often contentious) understanding of how the term *culture* can be defined. For instance, Kroeber and Kluckholn (1952) identified 164 separable definitions of culture within the anthropological literature alone. Fortunately, the design and evaluation of warnings does not require all of the finely honed theoretical distinctions made by anthropologists. From a sociological perspective, culture is the aggregation of experiences, values, beliefs, and attitudes that are communicated by social groups (Hofstede, 1997). Consistent with the purposes of this book as a whole, "culture" within this chapter will follow the definition of cultural psychologists Goldberger and Veroff (1995) as being "a system of shared meanings that ... provide a common lens for perceiving and structuring reality for its members" (p. 11). Because a population often includes large numbers of people who share different cultures, subcultures often coexist within groupings such as national boundaries or communities. Subcultures can be defined using a variety of dimensions, but one of the most significant in terms of warning and risk communication is ethnicity.

According to Yinger (1994), membership in an ethnic group is defined by the following characteristics: (1) others in the society perceive the group members to be different, (2) members identify themselves as different, and (3) members participate in shared activities related to their perceived common origin or culture (p. 3). Moreover, ethnic groups are often defined in terms of national origin, race, language, and religion (Gudykunst and Kim, 1997). In the development of warnings, the need for understanding how people of different ethnicities will interact with safety-related information is critical because members of subcultures typically share many of the values of the culture, but they "also have some values that differ from the larger culture" (Gudykunst, 1998, p. 43). Thus, efforts to protect the safety of the public from potential hazards must consider the heterogeneity of the people who receive the warning.

To illustrate the need for better understanding of how cultural attributes might impact the design and evaluation of warnings, consider the following demographic trends within the United States. Recent data from the US Census Bureau (2009) indicates that the American population totals approximately 304 million and that the most populous ethnic minority groups include those reporting Hispanic origin (15.4%), African Americans (12.9%), and Asians (4.5%). Population estimates indicate that by 2015, the number of those reporting Hispanic origin will increase to more than 57 million, the number of African Americans will increase to more than 42 million, and the number of Asians will increase to more than 16.5 million (U.S. Census, 2008). Thus, the ability to inform and protect all subgroups and ethnicities within our culture is dependent on understanding how these cultural attributes might affect warning effectiveness and related issues.

MODELING BEHAVIOR: HOW PEOPLE INTERACT WITH WARNINGS

A number of models could be used to serve as the basis of this discussion on warnings and culture (e.g., Edworthy and Adams, 1996; Lehto and Miller, 1986; Lindell and Perry, 2004; Rogers, Lamson, and Rousseau, 2000); however, the communication–human information processing (C-HIP) model described by Wogalter and associates (e.g., see Wogalter, 2006) provides a reasonable framework that is both comprehensive and consistent with the aforementioned persuasive communications models. In this chapter, C-HIP will be used to provide a theoretical framework for the discussion of cultural attributes. It is the context within which culture is discussed.

The C-HIP model has two major sections each with several component stages. A representation of the model can be seen in Figure 5.1. The first section of the framework uses some of the basic stages of a persuasive communication model (Hovland, Janis, and Kelley, 1953; Lasswell, 1948). To illustrate how these general

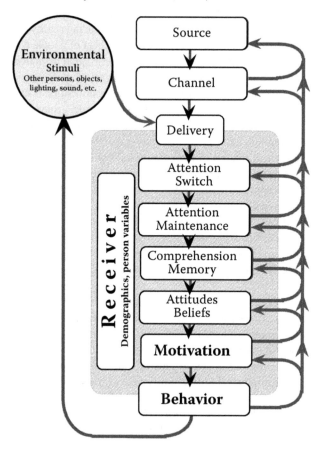

FIGURE 5.1   Communication-human information processing (C-HIP) model.

communication models can be altered to understand the warning process, McGuire (1980) provides a detailed description of communication theory with respect to warnings. Thus, the general framework for the C-HIP model focuses on a warning message being sent from one entity to another, that is, sent by a source (sender) through some channel(s) to a receiver. To place these C-HIP components within a cultural context, the Institute of Medicine (2002) suggests that cultural diversity should be considered when planning communication efforts by selecting credible sources, choosing message strategies, and determining channels for the delivery of safety information.

The second main section of the model focuses on the receiver and how people internally process information. This section interfaces with the first through effective delivery of the warning to individuals who are part of the target audience. When warning information is delivered to the receiver, processing may be initiated, and if not blocked in some way, will continue across several stages: from attention switch, attention maintenance, comprehension and memory, beliefs and attitudes, motivation, and possibly ending in behavior. Cultural attributes can be considered as an individual difference variable because each person who receives a warning belongs to a particular culture, and this varies from one individual to the next because the population is heterogeneous and diverse. The cultural aspect can be expected to operate at all levels of information processing within the receiver.

C-HIP MODEL

The C-HIP model is both a stage model and a process model. The C-HIP model is useful in describing a general sequencing of stages and the effects warning information might have as it is processed. If information is successfully processed at a given stage, the information "flows through" to the next stage. If processing at a stage is unsuccessful, it can produce a bottleneck, blocking the flow of information from getting to the next stage. If a person does not initially notice or attend to a warning, then processing of the warning goes no further. However, even if a warning is noticed and attended to, the individual may not understand it, and as a consequence, no additional processing occurs beyond that point. Even if the message is understood, it still might not be believed, thereby causing a blockage to occur at this point. If the person believes the message but is not motivated (to carry out the warning's instructed behavior), then the final stage involving compliance behavior might not occur. Successful processing in all stages results in safety compliance. While the processing of the warning may not make it all of the way to the behavioral compliance stage, it can still be effective at earlier stages. For example, a warning might enhance understanding and beliefs but not change behavior. While there are other aspects of the model (e.g., feedback loops), this basic model and its organization serves to provide a framework for our discussion of culture and warnings.

In the sections following, factors affecting each stage of the C-HIP model are described. The first three sections concern the section of C-HIP concerning communication from the source via some channel(s) to the receiver. Later sections concern analysis of information processing factors that are internal to the receiver.

SOURCE

A warning source is the entity or agency responsible for initiating hazard communication with the public. Sources can be government authorities, product manufacturers, media figures, or peers such as friends and relatives (Lindell and Perry, 2004; Wogalter, 2006). When an individual first encounters a warning, he or she judges the credibility of the source. Warnings originating from credible sources are likely to promote warning compliance, whereas less credible sources are likely to prompt information seeking. This process is known as warning confirmation and entails seeking information from other warning messages and different sources (Danzig, Thayer, and Galater, 1958). Because credibility varies between individuals, some have suggested that environmental warnings may be more believable to a larger segment of the population if they come from a mixed panel of scientists, public officials, reputable organizations, and familiar persons (Drabek and Stephenson, 1971). In fact, people are more likely to pay attention to warnings when they perceive that the source of information is "in the same boat" that they are; thus, shared involvement between the source and the receiver is likely to enhance risk perception (Aldoory and Van Dyke, 2006). Likewise, Weinstein's (1988) precaution adoption model suggests that the realization that a problem affects others "like you" can stimulate people to think about hazards and might lead them to plan to take preventative action by complying with a warning.

As source credibility is inherently tied to the concept of "trust," it is not surprising that trust is a topic of considerable discussion with no universally accepted scholarly definition (Rousseau, Sitkin, Burt, and Camerer, 1998). Although definitions of trust vary from one academic discipline to another, one finding from a growing body of research is particularly robust: trust and message credibility varies quite significantly by racial and ethnic status (Spence, Lachlan, and Griffin, 2007). For instance, African Americans frequently cite a distrust of government institutions and describe incidents of past exploitation such as the Tuskegee syphilis trials or Hurricane Katrina as explanations for an unwillingness to attend to or believe messages (Andrulis, Siddiqui, and Gantner, 2007; Freimuth et al., 2001). Likewise, differences in warning information exchange and dissemination have been observed between Mexican Americans, Caucasian Americans, and African Americans (Fothergill, Maestas, and Darlington, 1999).

To combat these deleterious effects of trust, obtaining community engagement during warning development is essential (Palenchar and Heath, 2007). Generally, such efforts have been described as one option for underrepresented segments of the population to take "community control" in an effort to counterbalance the power of the majority (Hacker, 1995). Thus, engaging participatory techniques that entail active collaboration between communities and other stakeholders such as government entities and aid organizations should provide a means of achieving this goal (George, Green, and Daniel, 1996). For instance, the formation of a community advisory board that includes faith-based organizations, community leaders, and community-outreach workers might be predicted to be useful in facilitating emergency risk communications such as warnings (Andrulis et al., 2007; Vaughan and Tinker, 2009). This approach where community leaders in refugee camps acted as part of an early warning system was demonstrated to be useful in preventing the outbreak of infectious diseases in Darfur (Pinto et al., 2005). Moreover, the interaction between credibility and source is further supported by research that suggests that local sources such as friends, family, and local news media might be considered "insider influences," which can be trusted more than "outsider influences," such as federal government entities or environmental groups (Baxter, 2009; Riley, Newby, and Leal-Almeraz, 2006).

CHANNEL

Warning channels refer to the communications medium used to transmit hazard information. Warnings can be transmitted in many ways. For instance, product warnings can be presented on labels directly on the product, on containers, in product manuals or inserts, on posters/placards, in brochures, and as part of audio–video presentations on various media (e.g., DVD or Internet). By contrast, environmental warnings might be disseminated via face-to-face contact, telephone, siren, radio, newspapers, television, and the Internet (e.g., Facebook, Twitter). Most commonly, warnings of either type use the visual (text and symbols) and auditory (alarms and voice) modalities as opposed to the other senses. There are exceptions, for example, an odor added to petroleum-based gases to enable detection by the olfactory sense, and the rough vibration of a product that is not mechanically functioning well can provide tactual, kinesthetic, and haptic sensation (Mazis and Morris, 1999; Cohen, Cohen, Mendat, and Wogalter, 2006).

Each of these channels varies in terms of the precision of dissemination and the specificity of the message (Lindell and Perry, 1987). For instance, a television or radio broadcast containing a flood warning might quickly reach the intended at-risk segment of the population, but dissemination is imprecise because the reception area for the station is larger than the risk area such that others who are not at risk will also receive the hazard information and erroneously believe themselves to be at risk. Also, recent evidence suggests that channel might interact with credibility such that incorrect information obtained from the Internet might be trusted, whereas correct information might be viewed with suspicion (Wogalter and Mayhorn, 2008). Face-to-face warnings can be much more targeted than mass media broadcasts. Given these shortcomings for each of the channels, it is often suggested that multiple channels be used to communicate with all members of society. For instance, recommendations regarding health communications about pandemic influenza suggest that authorities target several of the aforementioned channels as well as "ethnic radio and TV" (Vaughan and Tinker, 2009).

DELIVERY

While the source may try to disseminate warnings in one or more channels, the warnings might not reach some of the targets at risk (Williamson, 2006). Delivery refers to the point of reception where a warning arrives with the receiver. It is shown as a separate stage in the current C-HIP model shown in Figure 5.1 to emphasize its importance. A warning that a person sees or hears is a warning that has been delivered. Safety information provided on a DVD that is produced but never reaches the individual would be delivery failure. The reasons for failure to deliver the warning to targeted individuals can be multifold. The DVD may not have been distributed and sitting in bulk boxes in a warehouse. Or the distribution could be haphazard reaching some intended persons and not others. But even if individuals receive the video (e.g., via the Internet) they may not receive the needed information. For instance, groups with high rates of poverty may not have the playback equipment to see it or there might be a language barrier (e.g., limited proficiency in English). Of course, even if the person does see the video, it may not include the necessary warning. Thus, it may be necessary to distribute warning information in multiple ways to reach receivers at risk. The point is that if warnings given by a source do not reach the targets at risk, then the warning will have no or limited effects on the receiver.

Because technology is becoming ubiquitous in our society, the Internet is a constantly evolving channel for the delivery of safety information (Wogalter and Mayhorn, 2005). Although some portion of a population may have ready access to the Internet and frequently act in a proactive manner to search for information, others simply may not know that there is safety material (e.g., a list of recalled consumer products) that could be accessed. Thus, the existence of a digital divide must be recognized along with other disadvantages (and advantages) when Internet delivery is being considered as a mechanism for disseminating safety information to the public. Advantages might include the potential for timely, targeted, multimedia presentation of safety information that includes a gateway for further information seeking, whereas disadvantages might include the potential for inadvertently creating passivity as information is "pushed" to people thereby reducing interactivity with knowledgeable others (e.g., government officials). Ultimately, these advantages and disadvantaged need to be investigated via empirical research to determine whether the benefits exceed the costs in terms of safety.

ENVIRONMENTAL STIMULI

Besides the subject warning, other stimuli are usually simultaneously present. These stimuli may be other warnings or a wide assortment of nonwarning stimuli. These stimuli compete with the warning for the person's attention (described further below). With respect to a given warning, these other stimuli may be described as "noise" that could potentially interfere with warning processing. For example, a cellular telephone ringing or a baby crying just when an individual begins to examine a warning may cause distraction and lead to the warning not being fully read. The environment can have other effects. The illumination can be too dim to read the warning. In these cases of distraction or legibility, warnings of greater salience (e.g., light source added) could have better capability to attract and hold a person's focus.

Environmental influences often include other people as described in the social amplification of risk framework (Kasperson et al., 1988) that illustrates how interpersonal interactions in a social context can influence perception of risk. Awareness about what other people are doing in the local environment and elsewhere can affect warning compliance positively or negatively. As research by Masuda and Garvin (2006) illustrates, situated experiences of place can act as conflicting cultural worldviews that lead some individuals to act as risk amplifiers while others attenuate risk. For example, risk amplifiers might conclude that the risk of head injury is high, based on their observations that other people are wearing safety helmets on bicycles. Likewise, risk attenuators might conclude that the same risk is relatively low if they are surrounded by advertisements depicting people not wearing needed protective equipment, even though the product warning requires its use. Clearly then, the environment can have effects on warning processing. It shows a way of demonstrating or modeling ongoing processing. The source, receiver, other entities, and the environment can act on the situation and change it.

RECEIVER

The receiver is the person(s) or target audience to whom the warning is directed.

For a warning to effectively communicate information and influence behavior, the warning must first be delivered. Then attention must be switched to it and maintained long enough for the receiver to extract the necessary information. Next, the warning must be understood and must concur with the receiver's existing beliefs and attitudes. Finally, the warning must motivate the receiver to perform the directed behavior. The next several sections are organized around these stages of information processing.

ATTENTION SWITCH

An effective warning must initially attract attention, and to do so, it needs to be sufficiently salient (conspicuous or prominent). Warnings typically have to compete with other stimuli in the environment for attention. Several design factors influence how well warnings may compete for attention (see Wogalter and Leonard, 1999; Wogalter and Vigilante, 2006).

Larger is generally better. Increasing the overall size of the warning, its print size and contrast, generally facilitates warning conspicuousness. Context also plays an important role. It is not just the absolute size of the warning, but also its size relative to other displayed information. Color is an important attribute that can facilitate attention attraction (Bzostek and Wogalter, 1999; Laughery, Young, Vaubel, and Brelsford, 1993). However, recent evidence suggests that the interpretation of hazard severity associated with color varies by culture such that Chinese participants differed significantly from participants in the United States when both were asked to rank order colors in terms of perceived hazards (Lesch, Rau, Zhao, and Liu, 2009). Beyond interpretation of colors and their semantic meanings, other evidence suggests that perception of colors may also vary across cultures (Hupka, Zaleski, Otto, Reidel, and Tarabrina, 1997). Moreover, other problems unrelated to culture such as the presence of color blindness in some individuals suggests that color alone should not be relied on to attract attention yet color remains a frequently used design component in warnings.

Warning standards often use color as one of several components of the signal word panel to attract attention Other design components in the signal word panel include an alert symbol, the triangle/exclamation point, and one of three hazard connoting signal words (DANGER, WARNING, and CAUTION). Context again can also play a role with respect to color as a salience feature. An orange warning on a product label located on an orange product will have relatively less salience than the same warning conveyed using a different color. The color should be distinctive in the environment in which it is placed.

Symbols can also be useful for capturing attention. One example already mentioned is the alert symbol (triangle enclosing an exclamation point) used in the signal word panel in ANSI Z535 (2002; Bzostek and Wogalter, 1999; Laughery, 1993). This symbol only serves as a general alert. Bzostek and Wogalter (1999) found results showing people were faster in locating a warning when it was accompied by an icon. Other kinds of symbols may be used to convey more specific information. This latter purpose is discussed in the comprehension section (discussed later), but the point here is that a graphic configuration can also benefit the attention switch stage.

ATTENTION MAINTENANCE

Individuals may notice the presence of a warning but not stop to examine it. A warning that is noticed but fails to maintain attention long enough for its content to be encoded might serve as being of very little direct value. Attention must be maintained on the message for some length of time to extract meaning from the material. During this process, the information is encoded or assimilated with existing knowledge in memory.

With brief text or symbols, the warning message may be grasped very quickly, sometimes maybe as fast as a glance. For longer, more complex warnings, attention must be held for a longer duration to acquire the information. So to maintain attention in these cases, the warning needs to have qualities that generate interest so that the person is willing to maintain attention to it instead of something else. The effort necessary to acquire the information should be reduced as much as possible. Thus, there is a desire to enable the information to be grasped as easily as possible. Some of the same design features that facilitate the switch of attention also help to maintain attention. For example, large print not only attracts attention, but it also tends to increase legibility, which makes the print easier to read.

People will more likely maintain attention if a warning is well designed (i.e., aesthetic) with respect to formatting and layout. Research with western cultures suggests that people generally prefer warnings that are in a list outline format as opposed to continuous prose text (Desaulniers, 1987). Also, text messages presented in all caps are worse than mixed-case text in glance legibility studies (Poulton, 1967) and centered-line formatting is worse than left justified text (Hooper and Hannafin, 1986). Moreover, visual warnings formatted with plenty of white space and containing organized information groupings are more likely to hold attention than a single chunk of dense text (Wogalter and Vigilante, 2003; 2006). Interestingly, the lack of research with diverse samples may limit the potential usability of such design guidelines. For instance, the recommendations regarding the use of all caps may not be applicable to people who use pictoform languages such as Chinese, Japanese, or Korean. Likewise, suggestions regarding the use of left-justified text may not be applicable to readers of Arabic or Hebrew languages. Thus, there is an obvious need to test warning design features with other cultures.

Because individuals may decide it is too much effort to read large amounts of text, structured formatting could be beneficial in lessening the mental load and perception of difficulty. With perceptions of too much text, many prefer to direct their attention to something else. Formatting can make the visual display aesthetically pleasing to help hold people's attention on the material. Formatting can help process the information by "chunking" it into smaller units. Formatting can also show the structure or organization of the material, making it easier to search for and assimilate the information into existing knowledge and memory (Hartley, 1994; Shaver and Wogalter, 2003). Again, these recommendations are the result of very limited testing with homogeneous samples, and there is no guarantee that information will be processed similarly across cultures. Even if information processing is similar, research using the Cultural Sensitivity Assessment Tool to evaluate health-related information regarding cancer that targets African Americans suggests that readability is often reduced for these groups because efforts to use formatting and visual presentation are consistently underdeveloped (Guidry, Fagan, and Walker, 1998).

COMPREHENSION AND MEMORY

Comprehension concerns understanding the meaning of something, in this case, the intended message of the warning. Comprehension may derive from several components: subjective understanding such as its hazard connotation, understanding of language and symbols, and an interplay with the individual's background knowledge. Background knowledge is relatively permanent long-term memory structure that people carry with them. The sections below contain short reviews of some major conceptual research areas with respect to warnings and the comprehension stage. Again, much of this information is derived from limited testing that has not been validated across cultures; therefore, this section might be considered a set of "lessons learned" in investigating the use of various components of warning messages written in English.

Signal Words

Aspects of a warning can convey a level of subjective hazard to the recipient. The ANSI (2002) Z535 standard recommends three signal words to denote decreasing levels of hazard when US English is the language of the warning: DANGER, WARNING, or CAUTION (see also FMC Corporation, 1985; Peckham, 2006; Westinghouse Electric Corporation, 1981). The DANGER panel should be used when serious injury or death *will* occur if the directive is not followed. A WARNING panel is used when serious injury or death *may* occur if the directive is not followed. The CAUTION panel is used when less severe personal injuries or property damage may occur if the directive is not followed. While the standard describes CAUTION and WARNING with different definitions, numerous empirical research studies indicate that people do not readily distinguish between the two. The term DEADLY has been shown in several research studies to connote significantly higher hazard than DANGER (e.g., see Hellier and Edworthy, 2006; Wogalter, Kalsher, Frederick, Magurno, and Brewster, 1998; Wogalter and Silver, 1990, 1995).

While these general recommendations made in the ANSI standard (2002) are often used to construct safety messages for warning recipients within the United

States, cross-cultural safety research involving international populations suggests that differences in comprehension of signal word and color combinations might exist (Lesch et al., 2009). For instance, Lesch et al. (2009) found that US participants provided significantly higher mean ratings of perceived hazards to signal words than did the Chinese participants. Interestingly, other evidence suggests that hazard connotations assigned to colors and signal words might also vary between English-only and Spanish-speaking participants; therefore, warning designers within the United States might also exercise caution by examining the effects of culture (Wogalter, Frederick, Herrera, and Magurno, 1997).

Message Content

The content of the warning message should include information about the hazard, instructions on how to avoid the hazard, and the potential consequences if the hazard is not avoided (Wogalter, Godfrey, Fontenelle, Desaulniers, Rothstein, and Laughery, 1987).

a. *Hazard information.* At a minimum, the warning should identify the safety problem. Often, however, warnings might require more information regarding the nature of the hazard and the mechanisms that produce it.

b. *Instructions.* Warnings should instruct people about what to do or not do. The instructions should be specific inasmuch as reasonable to tell what exactly should be done or avoided. A classic nonexplicit warning statement is "Use with adequate ventilation." Two others are "May be hazardous to health" or "Maintain your tire pressure." These statements are inadequate by themselves to apprise people what they should or should not do. In the case of the statement "inadequate ventilation," does it mean to open a window, two windows, use a fan, or something more technical in terms of volume of airflow per unit time? In each case, without more information, users are left making inferences that may be partly or wholly incorrect (Laughery and Paige-Smith, 2006; Laughery, Vaubel, Young, Brelsford, and Rowe, 1993). Clearly, the use of certain terminology will be dependent on the language of the target audience. For instance, speakers of American or Canadian English are likely to recognize the term *truck* and make appropriate inferences, whereas speakers of British English, being more familiar with the term *lorry*, may not.

c. *Consequences.* Consequences information concerns what could result. It is not always necessary to state the consequences. However, one should be cautious in omitting it, because people may make the wrong inference. A common shortcoming of warnings is that the consequences information is not explicit, that is, it is lacking important specific details (Laughery and Paige-Smith, 2006; Laughery et al., 1993). The statement "May be hazardous to your health" in the context of an invisible radiation hazard is insufficient by itself as it does not tell what kind of health problem could occur. The reader could believe it could lead to minor burns not thinking that it could be something more severe, like cancer and perhaps death. In a later section, the telling of severe consequences is discussed as a factor in motivating compliance behavior.

The information contained in a warning message is also likely to influence public perception of situational risk associated with a particular hazard. Although much research has been conducted with receivers who speak English, it remains unclear whether such results (as illustrated below) can be generalized to other populations. With this caveat in mind, warning message content generally represents a source's assessment of the existence and seriousness of a threat as well as what the public should do to protect themselves (Lindell and Perry, 2004). Stylistic considerations governing the communication of warning content in English include certainty and clarity. Simply worded warning messages understandable to the public should be delivered with a high degree of certainty concerning the likelihood of hazard occurrence and the need to take preventative action (Perry, Lindell, and Greene, 1982). When message content is specific, warning recipients are likely to believe that the threat is credible and to personalize the risk that increases the likelihood that they will take some preventative action (Drabek and Stephenson, 1971). To illustrate, 80% of the approximately one million residents of New Orleans evacuated safely once they encountered dramatically worded warning messages that used strong statements such as "The area will be uninhabitable for weeks" and "Water shortages will make human suffering incredible by modern standards" (McCallum and Heming, 2006). Although the forecast and warning components of Hurricane Katrina have been described as well constructed, the post-Katrina relief and aid efforts were shameful in that they exposed complex societal issues linked to culture. For instance, even though the warnings were excellent, African Americans and those with a lower socioeconomic status were later identified as being particularly vulnerable to this disaster because they lacked the resources to evacuate. This instance clearly illustrates that just because a warning may work for one culture or income group it may not be applicable to others.

Symbols

Safety symbols may also be used to communicate the above-mentioned information in lieu of or in conjunction with text statements (e.g., Dewar, 1999; Mayhorn and Goldsworthy, 2007; Mayhorn and Goldsworthy, 2009; Mayhorn, Wogalter, and Bell, 2004; Wolff and Wogalter, 1998; Young and Wogalter, 1990; Zwaga and Easterby, 1984). Potentially, they can contribute to understanding when illiterates or nonreaders of the primary language are part of the target audience.

Comprehension is important for effective safety symbols (Dewar, 1999). Symbols that directly represent concepts are preferred because they are usually better comprehended than more abstract symbols (Magurno, Wogalter, Kohake, and Wolff, 1994; Wogalter, Silver, Leonard, and Zaikina, 2006; Wolff and Wogalter, 1993). Less directly represented concepts cannot always be developed, but with abstract and arbitrary symbols (Lesch, 2004; Wogalter, Sojourner, and Brelsford, 1997), the meaning has to be learned via training. Despite these apparent potential benefits to using symbols to convey hazard information, there have been a number of studies that show cultural differences in how people interpret the meaning of symbols. One example of such cultural differences was documented by Casey (1993) when he described a case report of Kurd villagers in northern Iraq. A skull and crossbones symbol was prominently displayed on containers of grain intended only for planting but not eating. Despite seeing the symbol, some Kurd villagers consumed the grain and became seriously ill because they thought that the picture of the skull and cross-bones was just a logo of some company.

Interestingly, cultural differences in symbol comprehension have been well documented by other researchers as well. When ANSI symbols were tested for comprehension in Ghana, severe interpretation discrepancies were noted for a number of symbols and their intended meanings (Smith-Jackson and Essuman-Johnson, 2002). Other research found that drivers from Canada, Israel, Finland, and Poland displayed large comprehension differences with traffic signs (Shinar, Dewar, Summala, and Zakowska, 2003). As already mentioned, Chinese and US participants varied in their interpretation of perceived hazards in a variety of warning component configurations. Likewise, residents of Hong Kong had difficulty interpreting the meaning of some safety signs used in mainland China (Chan and Ng, 2010). Thus, symbols should be tested for comprehension within the intended target audience (even when the perceived subcultures are geographically proximal to one another) prior to deployment in a public warning system.

Given these apparent cultural differences, it is important to assess safety symbol comprehension. What is an acceptable level of comprehension for safety symbols? Symbols should be designed to have the highest level of comprehension attainable; however, a quantitative metric would be useful to guide those tasked with developing such warning symbols. ISO 9186 (2001) provides comprehension criteria (see Deppa, 2006; Peckham, 2006) and specifies that testing should be conducted in at least three countries that vary by culture. Within the United States, the ANSI (2002) Z535 standard suggests a goal of at least 85% comprehension using a sample of 50 individuals representative from the target audience for a symbol to be used without accompanying text. If 85% cannot be achieved, the symbol may still have utility (e.g., for attention capture) as long as is not badly misinterpreted. According to the ANSI (2002) Z535 standard, an acceptable symbol within the United States must produce less than 5% critical confusions (opposite meaning or a meaning that would produce unsafe behavior). For instance, the pharmaceutical warning (see Figure 5.2)

Do Not

Get Pregnant

FIGURE 5.2   Accutane warning.

used on Accutane regarding the potential for birth defects if the substance is taken during pregnancy might be wrongly interpreted such that the text "Do Not Get Pregnant" in combination with the symbol (circle/slash image superimposed over a pregnant female body) means that the substance is for birth control (Mayhorn and Goldsworthy, 2007; 2009).

Level of Knowledge

The levels of knowledge and understanding of the warning recipients should be taken into consideration. Three cognitive characteristics of receivers that may vary by culture are important: language skill, reading ability, and technical knowledge.

In general, reading levels should be as low as feasible. For the general population in the United States, the reading level probably should be approximately the skill level of grades 4 to 6 (expected ability of 10- to 12-year-old readers), yet it should be recognized that other nations and cultures may utilize a different school system. Unfortunately, functional illiteracy pervades society on a worldwide scale. For example, in the United States, there are estimates of more than 16 million functionally illiterate adults. In other areas of the world such as Ghana, national literacy rates can be as low as 41% in rural areas (Ghana Statistical Service, 2000). If so, successful warning communication may require more than simply keeping reading levels to a minimum. The use of symbols, speech warnings, and special training programs may be beneficial adjuncts. Moreover, these potential methods may also benefit literate persons. A related consideration is that different subgroups within a population may speak and read different languages, or in other words, they are culturally different from the majority in a region or nation. Interestingly, measures of culture reveal remarkable diversity between geographic locations within relatively small regions (Hofstede, de Hilal, Malvezzi, Tanure, and Vinkin, 2010). Using the Hofstede Values Survey Module, these researchers found that one nation, in this case Brazil, could be decomposed into as many as five cultural regions that illustrated distinct differences due to the presence of Afro-Brazilian and indigenous Indian roots. Thus, these results suggest that an effective warning within a country must be able to cross cultural and language barriers. One such attempt within the United States was assessed by Lim and Wogalter (2003), who concluded that culturally inclusive warnings require the use of multiple languages, combined graphics, and transmission through multiple methods to reach various subpopulations that receive it.

BELIEFS AND ATTITUDES

Beliefs and attitudes is the next major stage of the C-HIP model, and it is here that cultural diversity plays an especially significant role in human information processing. As the classic work of Douglas and Wildavsky (1982) suggests, risk is a collective belief that is subject to cultural and social contexts. Beliefs refer to an individual's knowledge that is accepted as true (although some of it may not actually be true). It is related to the previous stage in that beliefs are formed from memory structure derived from social interactions with those who share their culture. Specifically, interpersonal interactions in a social context can influence perception of risk (Kasperson et al., 1988; Masuda and Garvin, 2006). In some respects, beliefs tend be more global and overarching compared to specific memories. An attitude is similar to a belief except it includes more affect or emotional involvement. Past research suggests that risk attitudes vary across culture (Smith-Jackson, 2006b). For instance, culture-specific fatalism, defined as the belief that safety outcomes are predetermined and externally controlled by others, was a powerful determinant of safety-related behavior in the Ivory Coast, West Africa (Kouabenan, 1998). More recently, Latino farmworkers reported higher risk perception associated with the use of pesticides and lower perceived control of their work environments than Americans of European descent (Smith-Jackson, Wogalter, and Quintela, 2010).

People's benign experiences with a potentially hazardous product can produce beliefs that a product is safer than it is. This quickly changes after being involved in some way with (or seeing) a serious injury event. According to the C-HIP model, a warning will be successfully processed at the beliefs and attitudes stage if the message concurs (or at least is not discrepant) with the receiver's current beliefs and attitudes. However, if the warning information does not concur, then beliefs and attitudes may need to be altered so that they concur before a person can have some motivation to carry out the warning's directed behavior. The message and/or other information needs to be persuasive to override existing incorrect beliefs and attitudes. Methods of persuasion are commonly used in advertising and have been empirically explored in the social and cognitive psychology literatures.

Perhaps one of the largest areas of research involves tailoring warning messages to meet the needs and capabilities of a specific target audience (Wogalter and Mayhorn, 2005). Efforts to engage in this use of persuasive messaging can be observed in the area of health-related communication. For instance, Uskul and Oysterman (2010) suggest that message frames or wording should be culturally salient and momentarily salient in convincing people to comply with persuasive safety messages. In this work, health communications were tailored to meet the cultural aspects of the audience members (i.e., Americans of European or Asian descent) to create self-relevance, termed *cultural salience*, whereas delivery of the matched messages following presentation of culturally relevant themes made the messages situationally relevant or "momentarily salient." To create these message characteristics, this research relied heavily on the cultural distinction that suggests that western cultures tend to possess an individualistic orientation that focuses on individual achievements and independent decision making, whereas eastern cultures tend to be collectivist cultures that value group relationships (Han and Shavitt, 1994; Triandis, 1995). Consistent with this concept, Uskul and Oysterman (2010) found that European Americans found individualistic message frames more persuasive than collectivist message frames, yet the opposite trend was true for Asian Americans. Further evidence suggests message tailoring can be used to alter antitobacco advertising in terms of theme and language to specifically target bicultural Mexican American youth, thereby resulting in changes to tobacco-related attitudes that were found to be moderators for a behavioral decrease in smoking (Kelly, Comello, Stanley, and Gonzalez, 2010).

Two relevant and interrelated factors associated with the beliefs and attitudes stage are hazard perception and relevance (see DeJoy, 1999; Riley, 2006; Vredenburgh and Zackowitz, 2006). Investigations of hazard perception suggests that the greater the perceived hazard, the more responsive people will be to warnings, as in looking for, reading, and complying with them. The converse is also true. People are less likely to look for, read, or comply with a warning for products that they believe are low in hazard. For instance, poisonous substances such as mercury are frequently used during cultural and religious practices by Latino and Caribbean communities that practice Santeria (Riley, Newby, and Leal-Almeraz, 2006). Not surprisingly, many of these religious users and practitioners did not perceive the material as being hazardous. Because the health-related consequences of mercury exposure are often delayed following exposure, many people may not tie the hazard to the consequence. This is important because the level of perceived hazard is also closely tied to beliefs about injury severity. People that perceive a product to be hazardous are more likely to act cautiously when they understand that injuries can be severe (Wogalter, Young, Brelsford, and Barlow, 1999). In contrast to these environmental hazards, injury likelihood is a much less important factor in perceptions of risk or hazard for more mundane consumer products (Wogalter, Brelsford, Desaulniers, and Laughery, 1991; Wogalter, Brems, and Martin, 1993).

In such cases where perceived risk is low, it is especially important that warning recipients perceive that a safety message is being directed to them and that the warning content is applicable to them. If perceived as irrelevant, the individual may instead attribute the warning as being directed to others and not personally. For example, men may utilize pharmaceutical substances such as Propecia (for male pattern baldness) that might cause birth defects if pregnant female family members come into contact with this medication. Ideally, men should be made aware of this aspect yet they might not believe pregnancy warnings apply to them (Mayhorn and Goldsworthy, 2007, 2009). In this particular case, there is a failure of comprehension because men may not understand their role in preventing female family members from coming in contact with the drug. One way to counter this is to personalize the warning so that it gets directed to relevant users and conveys facts that indicate that it is relevant (Wogalter, Racicot, Kalsher, and Simpson, 1994). Similarly, efforts to make health-related information culturally specific via tailoring (based on individual levels of religiosity, collectivism, racial pride, and time orientation) has resulted in stimulating information processing for African-American women exposed to cancer prevention and screening information (Kreuter and Haughton, 2006).

MOTIVATION

Motivation energizes the individual to carry out an activity. Some of the main factors that can influence the motivation stage of the C-HIP model are cost of compliance, severity of injury, and social influence. These topics are discussed below.

Compliance generally requires that people take some action, and usually there are costs associated with doing so. When faced with a warning, people frequently consider what compliance will cost them in terms of resources such as money, time, and effort (Kalsher and Williams, 2006). When describing their failure to evacuate from Hurricane Charley in 2004, many elderly Americans stated that they had nowhere to evacuate to (social cost), and they lived on a fixed income and lacked the financial resources (e.g., car, money) to evacuate (Mayhorn and Watson, 2006). Likewise, many people often cite their fear of looters as a reason to ignore evacuation orders (Mayhorn and Watson, 2006; McCallum and Heming, 2006). Practical interventions that might be used to rectify these concerns by alleviating fears might include assurances of security from authority figures as well as heightened awareness of free shelters.

The costs of noncompliance can also exert a powerful influence on compliance motivation. With respect to warnings, a main cost for noncompliance is severe injury consequences. Previous research suggests that people report higher willingness to comply with warnings when they believe there is high probability for incurring a severe injury (e.g., Wogalter et al., 1991, 1993, 1999). In fact, cultural differences in motivation and compliance lessen if people are convinced that a warning is accurate and risk is high (Perry and Lindell, 1991). When archival data for three ethnicities (i.e., Caucasians, African Americans, and Mexican Americans) were evaluated for evacuation compliance following a hazardous chemical spill, ethnicity was not a predictor of motivation to engage in protective action behavior.

Another motivator is social influence (Wogalter, Allison, and McKenna, 1989; Edworthy and Dale, 2000). For instance, seeing others not comply lessens the likelihood of compliance. However, when people see others comply with a warning, they are more likely to comply themselves (Cox and Wogalter, 2006). Often, group compliance might be considered an essential component of healthcare interventions. Previous research also suggests that the development of culturally targeted smoking cessation programs is more effective than traditional 12-step smoking cessation programs with African-American smokers (Matthews, Sanchez-Johnson, and King, 2009).

BEHAVIOR

The last stage of the sequential process is for individuals to carry out the instructions for warning-directed safe behavior (Kalsher and Williams, 2006; Silver and Braun, 1999). Warnings do not always affect behavior because of processing failures at earlier stages. Most research in this area focuses on the factors that affect compliance likelihood.

Some researchers have used "intentions to comply" as the method of measurement as a proxy to behavioral measurement because it is usually quite difficult to conduct behavioral tests. The reasons include the following difficulties: (a) researchers cannot expose participants to real risks because of ethical and safety concerns; (b) events that could lead to injury are relatively rare; (c) the construction scenario must appear to have a believable risk, yet at the same time must be safe; and (d) conducting behavioral compliance research is costly in terms of time and effort. Nevertheless, actual compliance is an important criterion for determining which factors work better than others to boost warning effectiveness and, consequently, safety behavior. Additionally, many products are used inside homes where access to determine how the product was used and whether a warning was complied with is difficult. In the future, it is likely that virtual reality will play a role in allowing research to be conducted in simulated conditions that avoid some of the above problems (Duarte,

Rebelo, and Wogalter, 2010). Unfortunately, these tools are not in widespread use and may not yet be available to many other researchers interested in cultural ergonomics.

Below, the following section on teratogenic warnings serves as a case study to illustrate the current, commonly available methodology and analysis techniques that can be used to assess the affects of culture on warning exposure. Consistent with the definition of culture used by Goldberger and Veroff (1995), young adult women constitute a culture in the sense that they share demographic/physical characteristics that separate them from males and they possess a system of attitudes regarding their own reproductive health that might impact how they perceive risks posed by pharmaceutical products.

REFINING TERATOGEN WARNING SYMBOLS: A CASE STUDY IN INCLUSIVE WARNING DESIGN AND EVALUATION METHODOLOGY

Medications such as Accutane, Propecia, and Thalidomide are used to treat a variety of clinical conditions such as acne, male pattern baldness, and cancer yet they share teratogenic properties that are known to cause severe birth defects. These properties are so toxic that even brief exposure to these medications during pregnancy or prior to conception can cause significant harm to the fetus (Meadows, 2001; Perlman, Leach, Dominguez, Ruszkowski, and Rudy, 2001). One approach to mitigating this increased risk of accidental exposure to teratogenic substances is to improve warnings that appear on pharmaceutical labels.

Unfortunately, previous research conducted at the Centers for Disease Control and Prevention (CDC) suggests that the teratogen warning that appeared on Accutane (up until it was recalled from U.S. markets in 2009) may be confusing to those who encounter it (Daniel, Goldman, Lachenmayr, Erickson, and Moore, 2001). Illustrated in Figure 5.2, the warning consists of a symbol showing a circle and a slash mark superimposed over a graphic representation of a pregnant woman with the accompanying text "Do Not Get Pregnant." Results reported by Daniel and her colleagues indicated that only 21 percent of the women exposed to the current warning were able to correctly interpret it. Moreover, 27 percent of those tested misinterpreted the warning to mean that the medication was a form of birth control.

As addressed above, a well-established benefit associated with the use of symbols is that people who cannot understand printed text warnings might be able to take advantage of pictorial safety information. Given the increasing cultural diversity of the U.S. population, the use of pictorial safety symbols has the potential to be "culturally neutral" (Edworthy and Adams, 1996). Unfortunately, assumptions of cultural neutrality cannot be relied upon unless verified by empirical investigation.

Given the shortcomings of the warning, efforts to improve patient comprehension through iterative design were implemented. Using such a technique, prototype warnings should be developed and tested for comprehension with a sample of the at-risk population. Warnings that do not meet acceptable levels of comprehension should be redesigned based on feedback from earlier test participants and retested for comprehension in an iterative process (design, test, redesign, test, etc.) until a satisfactory level of comprehension is reached. To demonstrate and carry out the process, Goldsworthy and Kaplan (2006a) described a process where rapid proto-typing, expert review, and user-centered design techniques were utilized to develop alternate teratogen warnings. Later, a field trial solicited open-ended interpretation of six candidate symbols from 300 participants (Goldsworthy and Kaplan, 2006b). These initial findings were promising because they revealed that participants' abilities to correctly interpret the meanings of several of the alternate warnings exceeded that of the existing warning, with several candidates emerging as viable alternatives to the existing warning. The candidates were further refined based on the results and a second, larger-scale field study (N = 700) was conducted to further validate these alternative warnings (Mayhorn and Goldsworthy, 2007). Results indicated that two of the alternate symbols exceeded 85% comprehension, and none exceeded 5% critical confusion. Also, the same two alternate symbols consistently elicited accurate responding in terms of message interpretation, target audience, intended action, and perceived consequences of ignoring the warning.

While these findings are useful in illustrating how warnings and other risk communications might be designed and evaluated, a related topic includes efforts to target a specific audience for communications purposes. To this end, audience analysis is a recognized technique that has been used for identifying the appropriate people and subgroups within a population that receive a warning (Smith-Jackson, 2006b). The section below offers an illustration of analytical tools that can be used to accomplish this task.

AUDIENCE ANALYSIS USING LATENT CLASS ANALYSIS

It is well known that audiences vary by a wide range of characteristics—some obvious, others not. It has become increasingly common to examine message interpretation not only by whether audiences get it right, but by who is getting it more or less right. For instance, risk perceptions associated with pesticide warning labels was found to differ between two ethnicities of farmworkers. The likelihood of warning compliance was found to be higher for European-American farm workers than for Latino farmworkers (Smith-Jackson, Wogalter, and Quintela, 2010). Similarly, in a study that examined several possible birth defects warning labels among a diverse group of women of childbearing age, both accuracy of warning interpretation and warning preference varied significantly by participant characteristics (Goldsworthy and Kaplan, 2006a; Mayhorn and Goldsworthy, 2007). These analyses typically examine common audience characteristics, such as age, gender, race and/or ethnicity by using simplistic statistical analytical tools such as Chi-square or Fisher's Exact Test to determine whether "correctness" or rates of particular responses vary by those demographic characteristics.

Such analytical approaches are useful in providing more information than simple descriptive statistics regarding percentages of correctness or types of responses across a sample. However, other statistical tools can provide a richer picture of audience segmentation, especially, but not only, when the hazardous situation involves multiple informational or behavioral components, when a sizable number of beliefs might be implicated in engagement (or disengagement) in a particular hazardous action, or when a complex set of demographic characteristics is suggested by previous research or previous researcher experience. For instance, Lim and Wogalter (2003) found that the perceptions of lengthiness and print size varied when Spanish and English speakers assessed multilingual warnings. With the realization that it is not always possible to generate different warnings for all subgroups of the population, one methodological approach that may be useful in identifying pertinent receiver characteristics for those interested in cultural ergonomics is latent class analysis (LCA).

LCA is part of a broad class of analyses that also includes latent profile analysis, latent class growth analysis, latent transition analysis, growth mixture modeling, and general growth mixture modeling (Muthén, 2001). The common denominator in these analyses is that respondents are assumed to come from different populations or subpopulations rather than from a single uniform population of respondents; accordingly, this family of analyses attempts to estimate and account for group membership as part of the analytic process. In practice, LCA is a method of grouping respondents into homogeneous subgroups based on their responses to a measure of interest. Thus, behavior and attitudes rather demographic variables might offer a more precise description of culture and it pertains to safety-related contexts.

Research by Goldsworthy, Mayhorn, and Meade (2010) examined the prescription medication loaning and borrowing behavior of 700 participants for 13 hypothetical scenarios. Examination of item endorsement probabilities and odds-ratios for all items included in the LCA revealed four distinct classes of medication loaners/ borrowers. Class 1 members had extremely low probabilities of ever having loaned or borrowed medicine and were very unlikely to share or borrow medicine under any hypothetical circumstance. For this reason, this class was labeled "Abstainers."

Class 2 respondents were very likely to have loaned or borrowed prescription medicines in the past. All Class 2 members indicated that they would share a medicine if they received it from a family member. Members of this class were also highly likely to share when they had the same problem as the person with the medicine or already had a prescription but ran out or did not have it with them. They would also be likely to share or borrow if they had an emergency, could not afford to buy the medicine, or wanted to help a friend. Conversely, respondents in this class were far less likely to share or borrow medicine when they wanted to relax or feel good, had heard a lot about the medicine from commercials, or wanted something to help them sleep. They were evenly split on whether they would share or borrow medicine for pain. Because medication history indicated a high probability of having previously loaned or borrowed medicine and the pattern of endorsement indicated that sharing likely occurred (or would occur) for pragmatic, situation-specific reasons, this group was labeled "Pragmatic Frequent Sharers."

Class 3 respondents were evenly split in their probability of having loaned or borrowed medicine during the past. However the probabilities of endorsing hypothetical situations under which they would share or borrow were very high. That is, while Class 3 respondents were somewhat less likely than Class 2 respondents to indicate previous loaning or borrowing, they were more likely than members of all other classes to say that they would share in each situation (with the exception of "got it from a family member"). Class 3 respondents were not only likely to endorse pragmatic reasons for loaning/borrowing, but they were also likely to endorse sharing situations that have little to do with access: they would borrow medicine to relax or feel good, help them sleep, or for pain. The probability of endorsing these items was much higher for Class 3 than for any other class. Members of Class 3 were also far more likely than members of other classes to indicate they would share or borrow a prescription medication that they had heard about from advertisements. Given the somewhat lower frequency of actual reported loaning/borrowing but the high probability of loaning or borrowing in the future in both pragmatic and outcome-based situations, this group was labeled "At-Risk Sharers." The At-Risk Sharers were significantly more likely than the other three classes to report making less than $25,000/year, despite showing no differences in employment status. The At-Risk Sharers also had a higher percentage of respondents, indicating that they were Hispanic and spoke Spanish as their primary language.

Finally, Class 4 respondents were unlikely to have loaned or borrowed medicine in the past and were generally unlikely to share or borrow in the future. The low probability of having previously loaned clearly differentiates this class from Class 2, as do the generally lower probabilities of future sharing associated with the hypothetical scenarios. However, unlike Class 1 Abstainers, this group would be somewhat likely to share under some circumstances (e.g., emergencies). Class 4 was labeled "Emergency Sharers."

The identification of latent classes based on behaviors of interest to warnings researchers facilitates tailoring warning messages to specific groups that can improve the cultural sensitivity of warnings as described above. Such targeting could increase the effectiveness of these warnings thereby promoting safety behavior for all segments of the population. For example, in this study, four types of medication sharers were identified based on patterns of endorsement: Abstainers, Pragmatic Frequent Sharers, At-Risk Sharers, and Emergency Sharers. Because each of these groups demonstrates different medication loaning and borrowing behaviors, they are likely to respond in different ways to messages about medication sharing.

Efforts to tailor safety-related messages for At-Risk Sharers might include the following examples. Because At-Risk Sharers are less likely to have previously shared but are more likely to do so in a wider variety of circumstances than all other groups, they should be made aware of the wide range of issues associated with specific types of sharing. Interestingly, the results also confirmed previous findings that low-income and Hispanic individuals may be disproportionately at risk for engaging in risky sharing behaviors than are other individuals. Given the high representation of low-income and Hispanic individuals in the At-Risk class and the finding that At-Risk Sharers are more likely to share when having heard about a medicine in advertisements, it seems important to note that drug advertisement disclaimers about risks and side effects are usually presented verbally in English, without visual accompaniment. It is reasonable to presume that such verbal messages are not discerned, much less understood, by non-English speakers. Changing these messages to more clearly communicate the potential side effects may be an important step toward mitigating risk broadly as well as specifically within these groups.

CONCLUSIONS AND RECOMMENDATIONS

The preceding review of the warnings literature was organized around the C-HIP model (Wogalter, 2006) and demonstrated how cultural factors can impact safety-related information transmitted via risk communications. This model divides the processing of warning information into separate stages that must be successfully completed for compliance behavior to occur. A bottleneck at any given stage can hinder processing at subsequent stages. Feedback from later stages can affect processing at earlier stages. Moreover, culture can influence information processing and interaction with safety-related information at any of the stages described in C-HIP. The model is valuable in describing some the processes and organizing a large amount of research.

In this chapter, the C-HIP model was used to demonstrate the rather sizable gaps that exist in our knowledge of warning diverse populations. While a number of the examples from the literature review did not measure culture per se, they did illustrate how communicating with diverse populations can be challenging. Using C-HIP to provide context, a number of general recommendations can be made to inform the design and evaluation of culturally inclusive warnings.

IDENTIFYING THE TARGET AUDIENCE

Before a warning can be effectively targeted to a particular segment of the population, efforts at audience analysis should be conducted to gather information regarding past behavior as well as the many dimensions of culture, including ethnicity, gender, socioeconomic status, age, and literacy (Smith-Jackson, 2006b). Ethnographic research methods such as interviews and participant observation (Riley, Newby, and Leal-Almeraz, 2006) or focus groups (Mayhorn, Nichols, Rogers, and Fisk, 2004) can be used to gain insight into existing audience characteristics such as risk perception and attitudes regarding particular hazards. To verify that the targeted groups are vulnerable to injury, some recent efforts have used focus groups in combination with archival analysis of national injury databases (McLaughlin and Mayhorn, in press). It makes sense to understand whether a hazardous situation exists or is probable prior to taking the time and effort to generate a warning. If such injury databases already exist (and researchers can gain access to them) to confirm the existence of a safety-related problem, it should be possible to analyze for behavioral differences that exist by common audience characteristics (e.g., ethnicity, gender, and age) through the use of descriptive statistical tools or latent-class analysis as described by Goldsworthy, Mayhorn, and Meade (2010). It should be recognized that sometimes the absence of such informational databases does not necessarily mean that a warning is not needed. Moreover, not all researchers or warning designers around the world have access to or understand complex statistical analyses.

USING PARTICIPATORY DESIGN TECHNIQUES TO RECRUIT
PARTICIPANTS AND ENGAGE THE COMMUNITY

Because cultural factors may be particularly associated with source credibility and variables related to message delivery, it is important to gain the confidence and active participation of the members of the target audience (George, Green, and Daniel, 1996; Palenchar and Heath, 2007). Not only will this relationship be useful in recruiting participants for later warning evaluation efforts, but it will also be useful in engaging the community in safety-related issues. Participatory ergonomics is an approach that has been widely used to understand the preexisting knowledge and experience of those who comprise the target audience (Kuorinka, 1997; van Eerd et al., 2010), and this has been particularly useful in promoting "safety culture" (Bentley and Tappin, 2010). For instance, the formation of a community advisory board that includes faith-based organizations, community leaders, and community-outreach workers should be an effective means of communicating with the target audience and potentially recruiting research participants who represent this population of interest (Smith-Jackson, 2006b; Vaughan and Tinker, 2009). In effect, such efforts will allow safety practitioners to become a part of the credible "insider influences" that can be trusted, thereby enabling access to members of different cultures (Baxter, 2009; Riley, Newby, and Leal-Almeraz, 2006).

DEVELOPING AND EVALUATING THE WARNING CONTENT VIA ITERATIVE DESIGN

Once the characteristics and activities of the target audience are known from previous interactions with the target audience via consumer testing and interviews, efforts to develop the content of safety communications can begin. Using what is known about the message frames or wording combinations that are most culturally salient (and understandable/credible, etc.), warning content can be tailored to meet the needs of the target audience (Uskul and Oysterman, 2010). Prototype warnings should be developed and tested for comprehension with multiple samples such as different ethic and cultural subgroups of the target audience in an iterative fashion (design, test, redesign, test, etc.). Warnings that do not meet acceptable levels of comprehension should be redesigned based on feedback from earlier test participants and retested for comprehension until a satisfactory level of comprehension is reached (Goldsworthy and Kaplan, 2006a, 2006b; Mayhorn and Goldsworthy, 2007, 2009).

FOLLOW-UP EVALUATION AFTER WARNING DEPLOYMENT

Once a prototype warning has undergone the aforementioned iterative process and it has been deployed to the public, the job of a safety communications practitioner is *not yet* complete. Efforts should be made to conduct a follow-up evaluation of warning message comprehension using a diverse, random sample of the target audience. While ANSI (2002) specifies that a minimum of 50 participants and ISO (2001) specifies that participants should come from at least three different countries, pictorial symbol comprehension testing needs to be culturally inclusive; therefore, stratified sampling methods that consider ethnicity, gender, age, and literacy should be implemented (Smith-Jackson, 2006b).

CONCLUSION

Along with the realization that culture can interact with any of the stages of the model, C-HIP can also be a valuable tool in systematizing the assessment process to help determine why a warning is not effective for particular portions of the target audience. It can aid in pinpointing where the bottlenecks in processing may be occurring and suggest solutions to allow processing to continue to subsequent stages. Warning effectiveness testing can be performed using methods described in the previous research. Evaluations of the processing can be directed to any of the stages described in the C-HIP model: source, channel, environment, delivery, attention, comprehension, attitudes and beliefs, motivation, behavior, and receiver variables. In effect, the model can be used as an investigative tool to determine why a warning is inadequately carrying out its function. In this chapter, C-HIP was used as a framework to highlight existing gaps of knowledge associated with the affect of culture as a receiver characteristic during the warning process.

In closing, there is an increasing recognition that culture plays an important role in risk communication (Kreuter and McClure, 2004). While the discussion presented here was not meant to provide a comprehensive review on all the ways that culture could potentially influence warning compliance, it was meant to act as a primer to inform those interested in cultural ergonomics of existing methodological and analytical techniques that might be employed to develop inclusive warning systems. The goal was to provide direction for future warning development and research. While much empirical work remains to be done, the promise of more culturally sensitive warning systems should be effective in promoting safety for all members of the public.

REFERENCES

Aldoory, L., and Van Dyke, M. A. (2006). The roles of perceived "shared" involvement and information overload in understanding how audiences make meaning of news about bioterrorism. *Journalism and Mass Communication Quarterly, 83*(2), 346–361.

Andrulis, D. P., Siddiqui, N. J., and Gantner, J. L. (2007). Preparing racially and ethnically diverse communities for public health emergencies. *Health Affairs, 26*(5), 1269–1279.

ANSI (2002). *Accredited Standards Committee on Safety Signs and Colors. Z535.1-5*, National Electrical Manufacturers Association, Arlington, VA.

Baxter, J. (2009). A quantitative assessment of the insider/outsider dimension of the cultural theory of risk and place. *Journal of Risk Research, 12*(6), 771–791.

Bentley, T., and Tappin, D. (2010). Incorporating organizational safety culture within ergonomics practice. *Ergonomics, 53* (10), 1167–1174.

Bzostek, J. A., and Wogalter, M. S. (1999). Measuring visual search time for a product warning label as a function of icon, color, column, and vertical placement. *Proceedings of the Human Factors and Ergonomics Society, 43*, 888–892.

Casey, S. (1993). *Set Phasers on Stun: And Other True Tails of Design, Technology, and Human Error*. Santa Barbara, CA: Aegean.

Chan, A. H. S., and Ng, A. W. Y. (2010). Investigation of guessability of industrial safety signs: Effects of prospective-user factors and cognitive sign features. *International Journal of Industrial Ergonomics, 40* (6), 689–697.

Cohen, H. H., Cohen, J., Mendat, C. C., and Wogalter, M. S. (2006). Warning channel: Modality and media. In M. S. Wogalter (Ed.), *Handbook of Warnings*. Mahwah, NJ: Lawrence Erlbaum Associates (Boca Raton, FL: CRC Press), chap. 9: pp. 123–134.

Cox, E. P., III, and Wogalter, M. S. (2006). Warning source. In M. S. Wogalter (Ed.), *Handbook of Warnings*. Mahwah, NJ: Lawrence Erlbaum Associates (Boca Raton, FL: CRC Press), chap. 8: pp. 111–122.

Daniel, K., Goldman, K., Lachenmayr, S., Erickson, J., and Moore, C. (2001). Interpretations of a teratogen warning symbol. *Teratology, 64,* 148–153.

Danzig, E. R., Thayer, P. W., and Galater, L. R. (1958). *The Effects of a Threatening Rumor on a Disease Stricken Community (National Research Council Disaster Study No. 10),* Washington D.C.: National Academy of Sciences.

DeJoy, D. M. (1999). Beliefs and attitudes. In M. S. Wogalter, D. M. DeJoy, and K. R. Laughery (Eds.), *Warnings and Risk Communication.* London: Taylor & Francis, pp. 183–219.

Deppa, S. W. (2006). U.S. and international standards for safety symbols. In M. S. Wogalter (Ed.), *Handbook of Warnings.* Mahwah, NJ: Lawrence Erlbaum Associates (Boca Raton, FL: CRC Press), chap. 37: pp. 477–486.

Desaulniers, D. R. (1987). Layout, organization, and the effectiveness of consumer product warnings. *Proceedings of the Human Factors Society, 31,* 56–60.

deTurk, M. A., and Goldhaber, G. M. (1988). Consumers' information processing objects and effects of product warning. *Proceedings of the Human Factors Society, 32,* 445–449.

Dewar, R. (1999). Design and evaluation of graphic symbols. In H. J. G. Zwaga, T. Boersema, and H. C. M. Hoonhout (Eds.), *Visual Information for Everyday Use: Design and Research Perspectives.* London: Taylor & Francis, pp. 285–303.

Douglas, M., and Wildavsky, A. (1982). *Risk and Culture.* Berkeley, CA: University of California Press.

Drabek, T. E., and Stephenson, J. S. (1971). When disaster strikes. *Journal of Applied Social Psychology 1*(2), 187–203.

Duarte, E., Rebelo, F., and Wogalter, M. (2010). Virtual reality and its potential for evaluating warning compliance. *Human Factors and Ergonomics in Manufacturing and Service Industries,* 20(6), 526–537.

Edworthy, J., and Adams, A. (1996). *Warning Design: A Research Prospective.* London: Taylor & Francis.

Edworthy, J., and Dale, S. (2000). Extending knowledge of the effects of social influence in warning compliance. *Proceedings of the XIVth Triennial Congress of the International Ergonomics Association and 44th Annual Meeting of the Human Factors and Ergonomics Society.* Santa Monica, CA: Human Factors and Ergonomics Society, vol. 4, 107–110.

Edworthy, J., and Hellier, E. (2006). Complex nonverbal auditory signals and speech warnings. In M. S. Wogalter (Ed.), *Handbook of Warnings.* Mahwah, NJ: Lawrence Erlbaum Associates (Boca Raton, FL: CRC Press), chap. 15: pp. 199–220.

Flynn, J., Slovic, P., Mertz, C. K., and Carlisle, C. (1999). Public support for earthquake risk mitigation in Portland, Oregon. *Risk Analysis, 19*(2), 205–216.

FMC Corporation (1985). *Product Safety Sign and Label System,* FMC Corporation, Santa Clara, CA.

Fothergill, A., Maestas, E. G. M., and Darlington, J. D. (1999). Race, ethnicity and disasters in the United States: A review of the literature. *Disasters, 23*(2), 156–173.

Frascara, J. (2006). Typography and the visual design of warnings. In M. S. Wogalter (Ed.), *Handbook of Warnings.* Mahwah, NJ: Lawrence Erlbaum Associates (Boca Raton, FL: CRC Press), chap. 29: pp. 385–406.

Freimuth, V. S., Quinn, S. C., Thomas, S. B., Cole, G., Zook, E., and Duncan, T. (2001). African American's views on research and the Tuskegee syphilis study. *Social Science and Medicine, 52,* 797–808.

George, M. A., Green, L. W., and Daniel, M. (1996). Evolution and implications of P. A. R. for public health. *Promotion and Education, 3*(4), 6–10.

Ghana Statistical Service. (2000). *Ghana Living Standards Survey 4.* Accra, Ghana: Author.

Goldberger, N. R., and Veroff, J. B. (1995). *The Culture and Psychology Reader.* New York: New York University Press.

Goldhaber, G. M., and deTurck, M. A. (1988). Effects of consumer's familiarity with a product on attention and compliance with warnings. *Journal of Products Liability, 11,* 29–37.

Goldsworthy, R. C., and Kaplan, B. (2006a). Warning symbol development: A case study on teratogen symbol design and evaluation. In M. S. Wogalter (Ed.), *Handbook of Warnings*. Mahwah, NJ: Lawrence Erlbaum Associates, pp. 739–754.

Goldsworthy, R. C., and Kaplan, B. (2006b). Exploratory evaluation of several teratogen warning symbols. *Birth Defects Research. Part A, Clinical and Molecular Teratology, 76*(6), 453–460.

Goldsworthy, R. C., Mayhorn, C. B., and Meade, A. W, (2010). Warnings in manufacturing: Improving hazard mitigation messaging through audience analysis. *Human Factors and Ergonomics in Manufacturing and Service Industries, 20* (6), 484–499.

Gudykunst, W. B. (1998). *Bridging Differences: Effective Intergroup Communication*. Thousand Oaks, CA: Sage.

Gudykunst, W. B., and Kim, Y. (1997). *Communicating with strangers*. New York: McGraw-Hill.

Guidry, J., Fagan, P., and Walker, V. (1998). Cultural sensitivity and readability of breats and prostate cancer education materials targeting African Americans. *Journal of the National Medical Association, 90*, 165–169.

Hacker, A. (1995). *Two Nations: Black and White, Separate, Hostile, Unequal*. New York: Ballantine Books.

Han, S., and Shavitt, S. (1994). Persuasion and culture: Advertising appeals in individualistic and collectivistic societies. *Journal of Experimental Social Psychology, 30*, 326–350.

Hartley, J. (1994). *Designing Instructional Text* (3rd ed.). London: Kogan Page/East Brunswick, NJ: Nichols.

Hellier, E., and Edworthy, J. (2006). Signal words. In M. S. Wogalter (Ed.), *Handbook of Warnings*. Mahwah, NJ: Lawrence Erlbaum Associates (Boca Raton, FL: CRC Press), chap. 30: pp. 407–417.

Hofstede, G. (1997). *Cultures and Organizations: Software of the Mind*. New York: McGraw-Hill.

Hofstede, G., de Hilal, A. V. G., Malvezzi, S., Tanure, B., and Vinken, H. (2010). Comparing regional cultures within a country: Lessons from Brazil. *Journal of Cross-Cultural Psychology, 41* (3), 336–352.

Hooper, S., and Hannafin, M. J. (1986). Variables affecting the legibility of computer generated text. *Journal of Instructional Development, 9*, 22–28.

Hovland, C., Janis, I., and Kelley, H. (1953). *Communication and Persuasion*. New Haven, CT: Yale University Press.

Hupka, R. B., Zaleski, Z., Otto, J., Reidl, L., and Tarabrina, N. V. (1997). The colors of anger, envy, fear, and jealousy: A cross-cultural study. *Journal of Cross-Cultural Psychology, 28*, 156–171.

Institute of Medicine (2002). *Speaking of Health: Assessing Health Communication Strategies for Diverse Populations*. Washington, DC: National Academy Press.

ISO (2001). *Graphical Symbols–Test Methods for Judged Comprehensibility and for Comprehension, ISO 9186*, International Organization for Standards.

Kalsher, M. J., and Williams, K. J. (2006). Behavioral compliance: Theory, methodology, and results. In M. S. Wogalter (Ed.), *Handbook of Warnings*. Mahwah, NJ: Lawrence Erlbaum Associates (Boca Raton, FL: CRC Press), chap. 21: pp. 289–300.

Kasperson, R. E., Renn, O., Slovic, P., Brown, H. S., Emel, J., Goble, R., Kasperson, J. X., and Ratick, S. (1988). The social amplification of risk: A conceptual framework. *Risk Analysis, 8*(2), 177–187.

Kelly, K., Comello, M. L. G., Stanley, L. R., and Gonzalez, G. R. (2010). The power of theme and language in multi-cultural communities which tobacco cessation messages are most persuasive to Mexican-American youth. *Journal of Advertising Research, 50* (3), 265–278.

Kouabenan, D. R. (1998). Beliefs and the perceptions of risks and accidents. *Risk Analysis, 18*, 243–252.

Kreuter, M. W., and Haughton, L. T. (2006). Integrating culture into health information for African American women. *American Behavioral Scientist, 49*(6), 794–811.

Kreuter, M. W., and McClure, S. M. (2004). The role of culture in health communication. *Annual Review of Public Health, 25*, 439–455.

Kroeber, A., and Kluckholn, C. (1952). *Culture.* New York: Random House.

Kuorinka, I. (1997). Tools and means of implementing participatory ergonomics. *International Journal of Industrial Ergonomics, 19*, 267–270.

Lasswell, H. (1948). The structure and function of communication in society. In L. Bryson (Ed.), *The Communication of Ideas.* New York: Harper, pp. 32–51.

Laughery, K. R. (1993). Everybody knows: Or do they? *Ergonomics in Design*, July, 8–13.

Laughery, K. R., and Paige-Smith, D. (2006). Explicit information in warnings. In M. S. Wogalter (Ed.), *Handbook of Warnings.* Mahwah, NJ: Lawrence Erlbaum Associates (Boca Raton, FL: CRC Press), chap. 31: pp. 419–428.

Laughery, K. R., Young, S. L., Vaubel, K. P., and Brelsford, J. W. (1993). The noticeability of warnings on alcoholic beverage containers. *Journal of Public Policy and Marketing, 12*, 38–56.

Lehto, M. R., and Miller, J. M. (1986). *Warnings: Volume 1. Fundamentals, Design and Evaluation Methodologies.* Ann Arbor, MI: Fuller Technical Publications.

Lesch, M. F. (2004). Comprehension and memory for warning symbols: Age-related differences and impact of training. *Journal of Safety Research, 34*, 495–505.

Lesch, M. F., Rau, P. P., Zhao, Z., and Liu, C. Y. (2009). A cross-cultural comparison of perceived hazard in response to warning components and configurations: US vs. China. *Applied Ergonomics, 40*, 953–961.

Lim, R. W., and Wogalter, M. S. (2003). Beliefs about bilingual labels on consumer products. *Proceedings of the Human Factors and Ergonomics Society, 47*, 839–843.

Lindell, M. K., and Perry, R. W. (2004). *Communicating Environmental Risk in Multiethnic Communities.* Thousand Oaks, CA: Sage Publications.

Lindell, M. K., and Perry, R. W. (1987). Warning mechanisms in emergency response systems. *International Journal of Mass Emergencies and Disasters, 5*, 137–153.

Magurno, A., Wogalter, M. S., Kohake, J., and Wolff, J. S. (1994). Iterative test and development of pharmaceutical pictorials. *Proceedings of the 12th Triennial Congress of the International Ergonomics Association, Vol 4, 360–362.*

Masuda, J. R., and Garvin, T. (2006). Place, culture, and the social amplification of risk. *Risk Analysis, 26* (2), 437–454.

Matthews, A. K., Sanchez-Johnson, L., and King, A. (2009). Development of a culturally targeted smoking cessation intervention for African American smokers. *Journal of Community Health, 34*(6), 480–492.

Mayhorn, C. B., and Goldsworthy, R. C. (2009). "New and improved": The role text augmentation and the application of responses interpretation standards (coding schemes) in a final iteration of birth defects warnings development. *Birth Defects Research Part A: Clinical and Molecular Teratology, 85*(10), 864–871.

Mayhorn, C. B., and Goldsworthy, R. C. (2007). Refining teratogen warning symbols for diverse populations. *Birth Defects Research Part A: Clinical and Molecular Teratology, 79*(6), 494–506.

Mayhorn, C. B., Nichols, T. A., Rogers, W. A., and Fisk, A. D. (2004). Hazards in the home: Using older adults' perceptions to inform warning design. *Journal of Injury Control and Safety Promotion, 11*(4), 211–218.

Mayhorn, C. B., and Podany, K. I. (2006). Warnings and aging: Describing the receiver characteristics of older adults. In M. S. Wogalter (Ed.), *Handbook of Warnings.* Mahwah, NJ: Lawrence Erlbaum Associates (Boca Raton, FL: CRC Press), chap. 26: pp. 355–362.

Mayhorn, C. B., and Watson, A. M. (2006). Older adult decision making during hurricane hazard preparation: To evacuate or shelter-in-place. *Proceedings of the 16th World Congress of the International Ergonomics Association.* Maastricht, The Netherlands.

Mayhorn, C. B., Wogalter, M. S., and Bell, J. L. (2004). Are we ready? Misunderstanding homeland security safety symbols. *Ergonomics in Design, 12*(4), 6–14.

Mazis, M. B., and Morris, L. A. (1999). Channel. In M. S. Wogalter, D. M. DeJoy, and K. R. Laughery (Eds.), *Warnings and Risk Communication.* London: Taylor & Francis, chap. 6.

McCallum, E., and Heming, J. (2006). Hurricane Katrina: An environmental perspective. *Philosophical Transactions of the Royal Society, Series A, 364,* 2099–2115.

McGuire, W. J. (1980). The communication-persuasion model and health-risk labeling. In L. A. Morris, M. B. Mazis, and I. Barofsky (Eds), *Banbury Report 6: Product Labeling and Health Risks.* Cold Spring Harbor, New York: Cold Spring Harbor Laboratory, pp. 99–122.

McLaughlin, A. C., and Mayhorn, C. B. (In press). Avoiding harm on the farm: Human factors. *Gerontechnology.*

Meadows, M. (2001). The power of Accutane. The benefits and risks of a breakthrough acne drug. *FDA Consumer Magazine, 35*(2), 18–23.

Muthén, B. O. (2001). Latent variable mixture modeling. In G. A. Marcoulides and R. E. Schumacker (Eds.), *New Developments and Techniques in Structural Equation Modeling.* Mahwah, NJ: Lawrence Erlbaum Associates, pp. 1–34.

Palenchar, M. J., and Heath, R. L. (2006). Strategic risk communication: Adding value to society. *Public Relations Review, 33,* 120–129.

Peckham, G. M. (2006). ISO design standards for safety signs and labels. In M. S. Wogalter (Ed.), *Handbook of Warnings.* Mahwah, NJ: Lawrence Erlbaum Associates (Boca Raton, FL: CRC Press), chap. 21: pp. 455–462.

Perlman, S. E., Leach, E. E., Dominguez, L., Ruszkowski, A. M., and Rudy, S. J. (2001). "Be smart, be safe, be sure": The revised Pregnancy Prevention Program for women on isotretinoin. *Journal of Reproductive Medicine, 46*(2 Suppl.), 179–85.

Perry, R. W., Lindell, M. K., and Greene, M. R. (1982). Threat perception and public response to volcano hazard. *Journal of Social Psychology, 116,* 119–204.

Perry, R. W., and Lindell, M. K. (1991). The effects of ethnicity on evacuation decision-making. *Int. J. Mass Emerg. Disasters, 9,* 47–68.

Pinto, A., Saeed, M., El Sakka, H., Rashford, A., Colombo, A., Valenciano, M., and Sabatinelli, G. (2005). Setting up an early warning system for epidemic-prone diseases in Darfur: A participative approach. *Disasters, 29*(4), 310–322.

Poulton, E. (1967). Searching for newspaper headlines printed in capitals or lower-case letters. *Journal of Applied Psychology, 51,* 417–425.

Reid, P. T. (1995). Poor women in psychological research: shut up and shut out. In N. R. Goldberger and J. B. Veroff (Eds.), *The Culture and Psychology Reader.* New York: New York University Press, pp. 184–204.

Riley, D. M. (2006). Beliefs, attitudes, and motivation. In M. S. Wogalter (Ed.), *Handbook of Warnings.* Mahwah, NJ: Lawrence Erlbaum Associates (Boca Raton, FL: CRC Press), chap. 21: pp. 289–300.

Riley, D. M., Newby, C. A., and Leal-Almeraz, T. O. (2006). Incorporating ethnographic methods in multidisciplinary approaches to risk assessment and communication: Cultural and religious uses of mercury in Latino and Caribbean communities. *Risk Analysis, 26*(5), 1205–1221.

Rogers, W. A., Lamson, N., and Rousseau, G. K. (2000). Warning research: An integrative perspective. *Human Factors, 42,* 102–139.

Rousseau, D. M., Sitkin, S. B., Burt, R. S., and Camerer, C. (1998). Not so different after all: A cross-discipline view of trust. *Academy of Management Review, 23,* 393–404.

Rousseau, G. K., Lamson, N., and Rogers, W. A. (1998). Designing warnings to compensate for age-related changes in perceptual and cognitive abilities. *Psychology and Marketing*, 15(7), 643–662.

Shaver, E. F., and Wogalter, M. S. (2003). A comparison of older v. newer over-the-counter (OTC) nonprescription drug labels on search time accuracy. *Proceedings of the Human Factors and Ergonomics Society 47th Annual Meeting*, Santa Monica, CA: HFES.

Shinar, D., Dewar, R. E., Summala, H., and Zakowski, L. (2003). Traffic symbol comprehension: A cross-cultural study. *Ergonomics, 46*(15), 1549–1565.

Silver, N. C., and Braun, C. C. (1999). Behavior. In M. S. Wogalter, D. M. DeJoy, and K. R. Laughery (Eds.), *Warnings and Risk Communication.* London: Taylor & Francis, pp. 245–262.

Smith-Jackson, T. L. (2006a). Receiver characteristics. In M. S. Wogalter (Ed.), *Handbook of Warnings.* Mahwah, NJ: Lawrence Erlbaum Associates (Boca Raton, FL: CRC Press), chap. 24: pp. 335–344.

Smith-Jackson, T. L. (2006b). Culture and warnings. In M. S. Wogalter (Ed.), *Handbook of Warnings.* Mahwah, NJ: Lawrence Erlbaum Associates (Boca Raton, FL: CRC Press), chap. 27: pp. 363–372.

Smith-Jackson, T. L., and Essuman-Johnson, A. (2002). Cultural ergonomics in Ghana, West Africa: A descriptive study of industry and trade workers' interpretations of safety symbols. *International Journal of Occupational Safety and Ergonomics, 8*(1), 37–50.

Smith-Jackson, T., Wogalter, M. S., and Quintela, Y. (2010). Safety climate and risk communication disparities for pesticide safety in crop production by ethnic group. *Human Factors and Ergonomics in Manufacturing, 20*(6), 511–525.

Spence, P. R., Lachlan, K. A., and Griffin, D. R. (2007). Crisis communication, race, and natural disasters. *Journal of Black Studies, 37*(4), 539–554.

Triandis, H. C. (1995). *Individualism and Collectivism.* Boulder, CO: Westview Press.

U.S. Census Bureau (2009). *Monthly Resident Population Estimates by Age, Sex, Race, and Hispanic Origin for the United States.* Washington, DC: U.S. Government Printing Office.

U.S. Census Bureau (2008). *2008 National Population Projections.* Washington, DC: U.S. Government Printing Office.

Uskul, A. K., and Oysterman, D. (2010). When message-frame fits salient cultural-frame, messages feel more persuasive. *Psychology and Health, 25*(3), 321–337.

Van Eerd, D., Cole, D., Irvin, E., Mahood, Q., Keown, K., Theberge, N., Village, J., St Vincent, M., and Cullen, K. (2010). Process and implementation of participatory ergonomic interventions: A systematic review. *Ergonomics, 53* (10), 1153–1166.

Vaughan, E., and Tinker, T. (2009). Effective health risk communication about pandemic influenza for vulnerable populations. *American Journal of Public Health, 99*(S2), S324–S332.

Vredenburgh, A. G., and Zackowitz, I. B. (2006). Expectations. In M. S. Wogalter (Ed.), *Handbook of Warnings* (Chap. 25: pp. 345–354). Mahwah, NJ: Lawrence Erlbaum Associates (Boca Raton, FL: CRC Press).

Weinstein, N. (1988). The precaution adoption process. *Health Psychology, 7*, 355–386.

Westinghouse Electric Corporation (1981). *Product Safety Label Handbook.* Trafford, PA: Westinghouse Printing Division.

Williamson, R. B. (2006). Fire warnings. In M. S. Wogalter (Ed.), *Handbook of Warnings.* Mahwah, NJ: Lawrence Erlbaum Associates (Boca Raton, FL: CRC Press), chap. 56: pp. 701–710.

Wogalter, M. S. (2006). *Handbook of Warnings.* Mahwah, NJ: Lawrence Erlbaum Associates (Boca Raton, FL: CRC Press).

Wogalter, M. S., Allison, S. T., and McKenna, N. (1989). Effects of cost and social influence on warning compliance, *Human Factors*, vol. 31, pp. 133–140.

Wogalter, M. S., Brelsford, J. W., Desaulniers, D. R., and Laughery, K. R. (1991). Consumer product warnings: The role of hazard perception. *Journal of Safety Research, 22,* 71–82.

Wogalter, M. S., Brems, D. J., and Martin, E. G. (1993). Risk perception of common consumer products: Judgments of accident frequency and precautionary intent. *Journal of Safety Research, 24,* 97–106.

Wogalter, M. S., DeJoy, D. M., and Laughery, K. R. (Eds.). (1999). *Warnings and Risk Communication.* London: Taylor & Francis.

Wogalter, M. S., Frederick, O. L., Herrera, A. B., and Magurno, A. (1997). Connoted hazard of Spanish and English warning signal words, colors, and symbols by native Spanish language users. *Proceedings of the 13th Triennial Congress of the International Ergonomics Association,* IEA '97, 3, 353–355.

Wogalter, M. S., Godfrey, S. S., Fontenelle, G. A., Desaulniers, D. R., Rothstein, P. R., and Laughery, K. R. (1987). Effectiveness of warnings. *Human Factors, 29,* 599–612.

Wogalter, M. S., Kalsher, M. J., Frederick, L. J., Magurno, A. B., and Brewster, B. M. (1998). Hazard level perceptions of warning components and configurations. *International Journal of Cognitive Ergonomics, 2,* 123–143.

Wogalter, M. S., and Leonard, S. D. (1999). Attention capture and maintenance. In M. S. Wogalter, D. M. DeJoy, and K. R. Laughery (Eds.), *Warnings and Risk Communication.* London: Taylor & Francis, pp. 123–148.

Wogalter, M. S., and Mayhorn, C. B. (2008). Trusting the Internet: Cues affecting perceived credibility. *International Journal of Technology and Human Interaction, 4*(1), 76–94.

Wogalter, M. S., and Mayhorn, C. B. (2005). Providing cognitive support with technology-based warning systems. *Ergonomics, 48*(5), 522–533.

Wogalter, M. S., Racicot, B. M., Kalsher, M. J., and Simpson, S. N. (1994). The role of perceived relevance in behavioral compliance in personalized warning signs. *International Journal of Industrial Ergonomics, 14, 233–242.*

Wogalter, M. S., Silver, N. C., Leonard, S. D., and Zaikina, H. (2006). Warning symbols. In M. S. Wogalter (Ed.), *Handbook of Warnings.* Mahwah, NJ: Lawrence Erlbaum Associates (Boca Raton, FL: CRC Press), chap. 12: pp. 159–176.

Wogalter, M. S., Sojourner, R. J., and Brelsford, J. W. (1997). Comprehension and retention of safety pictorials. *Ergonomics, 40,* 531–542.

Wogalter, M. S., and Silver, N. C. (1995). Warning signal words: Connoted strength and understandability by children, elders, and non-native English speakers. *Ergonomics, 38,* 2188–2206.

Wogalter, M. S., and Silver, N. C. (1990). Arousal strength of signal words. *Forensic Reports, 3,* 407–420.

Wogalter, M. S., and Vigilante, W. J., Jr. (2003). Effects of label format on knowledge acquisition and perceived readability by younger and older adults. *Ergonomics, 46,* 327–344.

Wogalter, M. S., and Vigilante, W. J., Jr. (2006). Attention switch and maintenance. In M. S. Wogalter (Ed.), *Handbook of Warnings.* Mahwah, NJ: Lawrence Erlbaum Associates (Boca Raton, FL: CRC Press), chap. 18: pp. 245–266.

Wogalter, M. S., Young, S. L., Brelsford, J. W., and Barlow, T. (1999). The relative contribution of injury severity and likelihood information on hazard-risk judgments and warning compliance. *Journal of Safety Research, 30,* 151–162.

Wolff, J. S., and Wogalter, M. S. (1998). Comprehension of pictorial symbols: Effects of context and test method. *Human Factors, 40,* 173–186.

Young, S. L., Laughery, K. R., Wogalter, M. S., and Lovvoll, D. (1999). Receiver characteristics in safety communications. In W. Karwowski and W. S. Marras (Eds.), *The Occupational Ergonomics Handbook,* Boca Raton, FL: CRC Press, pp. 693–706.

Young, S. L., and Wogalter, M. S. (1990). Comprehension and memory of instruction manual warnings: Conspicuous print and pictorial icons. *Human Factors, 32,* 637–649.

Yinger, M. (1994). *Ethnicity.* Albany: State University of New York Press.

Zwaga, H. J. G., and Easterby, R. S. (1984). Developing effective symbols or public information. In R. S. Easterby and H. J. G. Zwaga (Eds.), *Information Design: The Design and Evaluation of Signs and Printed Material.* New York: John Wiley & Sons.

6 Cultural Ergonomics Perspectives on Occupational Safety and Health

Sharnnia Artis and Tonya Smith-Jackson

CONTENTS

INTRODUCTION

Occupational Safety and Health (OSH) is a cross-disciplinary field whose aim is to protect the safety, health, and welfare of people engaged in work or employment. With OSH covering a vast range of industries and occupations, and with its focus on a diverse set of workplace challenges (e.g., hazards and exposures, diseases and injuries, safety and prevention, chemicals, and emergency preparedness and response), safety and health in the workplace introduce a level of complexity that requires innovative approaches to prevention and control. Researchers, practitioners, and employers focus on establishing and maintaining a safe working environment for all workers. Thus, workplace safety has become extremely important in an era of escalating health care costs and an increasingly litigious response to on-the-job accidents and fatalities. When workplace safety is managed appropriately, employers often see a decline in organizational costs, but also in organizational reputation.

Companies such as DuPont and Caterpillar enjoy strong reputations as businesses that care about the safety and health of workers. But, as workers' backgrounds, cultures, national origins, and experiences continue to increase in diversity, knowing and understanding workers' safety and health needs, beliefs, attitudes, and practices will become increasingly difficult to assess, especially using the tools and methods that originated from Euro-centric researchers and OSH practitioners.

As the US workforce becomes increasingly diverse and global, attention is focused on disparities in worker safety and health across racial and ethnic populations. The rise in disparities has been attributed to both an overrepresentation of racial and ethnic minority workers in the most hazardous industries and a lack of culturally relevant interventions for certain worker populations due to barriers created by social, cultural, and economic factors, including language, literacy, and marginal economic status (CDC, 2011). While the pathways to which socioeconomic status and race/ethnicity influence OSH are complex and still not completely understood, one pathway receiving increased attention is the nature of work itself and how it can lead to social inequities in safety and health (Muntaner and Schoenbach, 1994). To ensure a work environment that yields equitable benefits for all workers, researchers, practitioners, and employers have to consider the cultural attributes of the workers and the environment, and must focus on the intersections of the person, work system, technology, and environment. Applying research-based cultural ergonomic approaches may lead to more cost-effective improvements that will benefit all workers, including majority-group members. In this chapter, the primary goal is to describe applications of cultural ergonomic approaches, interventions, and recommendations to the workplace, with a specific focus on construction, manufacturing, and crop production.

WORKERS AND CULTURAL DIFFERENCES

Across the OSH domains, there are workers comprising different ethnicities, socioeconomic statuses, genders, and age. In OSH, overlooking differences associated with these cultures may result in an increase in injuries and fatalities, and an increase in cost from litigation and workers' compensation insurance (as well as premiums). Researchers generally agree that individuals belonging to a specific ethnic or cultural group tend to share a common understanding of their own ethnicity or culture (Sasao and Sue, 1993). While workers are self-aware and hold in-group knowledge of their shared cultures, employers are seldom aware of the significance of these identifies to the safety, health, and success of the workplace.

In this section, we will discuss these cultural differences for ethnic groups and gender in greater detail. The discussion in this chapter leans toward a somewhat stronger focus on the United States, although it should be recognized that disparities and other differences impacting OSH apply to all nations. The focus on the United States is due, in part, to the expertise of the authors. However, it should also be noted that the diversity of the United States is often ignored within the United States and by other OSH researchers globally, who perceive the United States to be a homogeneous culture. A distinctly "USA" or "American" culture does not exist, or is at best, very difficult to define and bound; although some have represented that culture by using the popular media. The popular media in the United States tends to focus on only one small portion of the so-called American culture, while leaving a host of other cultures within the country unrecognized and unacknowledged, i.e., Appalachian culture, Southern culture, Chinese-American culture, Midwestern culture, inner-city culture, etc.

ETHNIC GROUPS

Compared to their Caucasian counterparts in the United States, African-American, Hispanic or Latino workers, and immigrant workers of other ethnicities (such as Eastern Europeans and individuals from Laos) are disproportionately employed in some of the most dangerous occupations. In this volume, Hispanic and Latino will be used in line with the US definition of Hispanic and Latino, meaning any individual who identifies as Cuban, Mexican, Puerto Rican, South or Central American, or any other Spanish culture or origin regardless of race (Ennis, Rios-Vargas, Albert, 2011). African-American males are twice as likely as White males to work in service occupations and as laborers, fabricators, and operators, yet are half as likely to be in managerial or professional specialty occupations. Some research by Daniels in 2004 revealed an injury rate that was about a third higher for both African-American male and female workers compared to White workers (Daniels, 2004). The over-representation in more hazardous occupations may account for some of the higher injury rates among African-American workers. In 2005, a recent Bureau of Labor Statistics (BLS) study reported similar trends for Hispanics. The study found that the fatal injury rate for non-Latino Whites and Blacks had steadily declined, while the fatal injury rate for Latinos increased. Loh and Richardson (2004) examined BLS data from 1996 through 2001 to identify current trends in fatal work injuries among foreign-born workers. Their study concluded that Mexican immigrants, in particular, are at higher risk for nonfatal workplace injuries or illnesses than any other race or ethnic group.

As noted earlier, many studies have concluded that Latino immigrants may be at greater risk for workplace injuries than their non-Latino counterparts, especially in the construction industry (Anderson, Huntington, and Welch, 2000; Dong and Platner, 2004; Loh and Richardson, 2004). A number of factors have been associated with the higher injury rates among Latinos. One important dynamic that has been routinely cited is the disproportionate representation of this ethnic group in higher-risk construction jobs (Anderson et al., 2000; Jackson and Loomis, 2002; Loh and Richardson, 2004). For example, from 1997 to 1999, the Texas Workers' Compensation Commission records indicated that 45.5% of the fatalities involved Latino workers. These workers shared similar characteristics, such as low-skill levels, being fairly young, working in hazardous and physically demanding occupations, and being foreign-born (Fabrego and Starkey, 2001).

According to US Census statistics, as of 2010, Latinos totaled 50.5 million persons or about 16 percent of the total population of the United States in year 2010 (Ennis, Rios-Vargas, and Albert, 2011). By 2050, Latinos are expected to compose 25 percent, or 96 million residents. Many of these Spanish-speaking workers come from less affluent countries where the majority of the population lives at a very low socioeconomic level. Although Latinos were always part of the United States and some territories previously owned by Latinos were taken by the United States, there is still a large number who enter the borders of the United States seeking better working conditions and better pay, and fleeing political oppression. Even without formal training and education in the English language, and due to a strong need to earn wages, immigrant workers may accept and tolerate dangerously high and unreasonable risks in the workplace. This pattern is similar to newly arrived immigrant workers who work in other affluent countries dominated by a majority and more affluent ethnic group.

GENDER

The workplace role of women in construction is still being defined. However, what is known are issues that should be considered when employing women in the workplace. Anthropometrics, the study of the human body dimensions, should not be overlooked. Men and women are proportioned differently. There is variation in sizes among racial and ethnic groups. Some solutions to accommodate these differences include training, providing proper equipment and facilities, and personal protective equipment that fits, is appropriate for the job, and is properly maintained (Abrams, 2006).

The *Roanoke Times* newspaper (2004) highlighted the accomplishments of the first African-American female to be accepted into the Bricklayers Union. Davis, who was in her 40s at the time, was away from the job for 6 months because of a fall from scaffolding at a height of 50 feet. She landed on a pile of bricks, and broke most of her ribs while sustaining a number of other serious injuries. She was back on the job as soon as her physician allowed. Davis, who lives in Kempsville, Virginia, reported that male bricklayers did not like sharing their jobs with a woman. She also described what it means to be an older minority woman in construction, suggesting that it was quite challenging. Interestingly, she attributed her success to the few male colleagues who accepted her and were willing to train her. Although the report given by Davis was broad, it alludes to the "culture" of construction and the tendency of some workers to try to sustain the male-gender-typed environment of construction. Unfortunately, the prejudices and desires to sustain the traditional attributes of an environment may put workers either directly or indirectly at risk for accidents.

Research on gender in the occupational setting reveals little evidence that gender stereotyping has abated or decreased. Diekman and Eagly (2000) and Holt and Ellis (1998) suggest that even in the face of more gender equivalence across occupations (more jobs with equal or near equal gender proportions), gender stereotypes still persist in the workplace. This delayed change in attitudes after actual environmental change via integration is referred to as "conservative lag" by sociologists (similar to the sluggish beta in signal detection theory). Tepper, Brown, and Hunt (1993) and Rudman and Glick (1999) found that women who were perceived as competitive-aggressive were rated lower in likability and mental stability. Violations of gender stereotypes by not displaying the "expected behaviors" can undermine group process and result in avoidance or punishments in the shared environment (Prentice and Carranza, 2004).

CULTURALLY RELEVANT RESEARCH IN OSH

SOCIOTECHNICAL SYSTEMS MODEL

Sociotechnical systems (STS) theory has been used for decades as a framework to design and understand organizations and to facilitate organizational changes (Hendrick, 1991). The concept of STS defines organizations as open systems engaged in transforming inputs into desired outputs. As seen in Figure 6.1, the STS framework divides an organization into four interdependent subsystems: (1) personnel, (2) technological, (3) organizational design, and (4) environment (Hendrick, 1991; Hendrick and Kleiner, 2001; Hendrick and Kleiner, 2002). The next sections will discuss the different subsystems of small construction firms in detail.

PERSONNEL SUBSYSTEM

The personnel subsystem corresponds to people in an organization. There are three important characteristics of the personnel subsystem that are sensitive to the design of a work system structure: (1) demographic characteristics, (2) degree of professionalism, and (3) psychosocial aspects of the workforce (Hendrick and Kleiner, 2002). Examples of demographic characteristics include the age of the workforce, the degree of cultural diversity in the workplace, and the gender breakdown of the workforce. The degree of professionalism refers to the extent to which a person has learned and accepted the values, norms, and expected behaviors of the job before accepting a position within an organization. The psychosocial aspects of the workforce include the personality of the organization's workforce, especially in terms of how receptive the workers are to new ideas and concepts. According to Taylor and Felten (1993), the personnel subsystem functions to (1) attain the system's primary goals, (2) adapt to the external environment for survival, (3) integrate internal environment for conflict management, and (4) provide for the development and

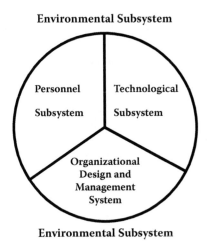

FIGURE 6.1   Sociotechnical subsystems framework.

maintenance of the system's long-term needs. Ideally, the workers should be intrinsically motivated to perform these functions on behalf of the organization. In reality, however, this motivation does not always exist within every worker, which causes the work system to perform at a suboptimal level.

ORGANIZATIONAL DESIGN SUBSYSTEM

An organization is a unit consisting of two or more people functioning on a relatively continuous basis and through a division of labor and an understood hierarchy of authority to achieve a common goal (Hendrick and Kleiner, 2001). Organizational design encompasses the design of an organization's work system, structure, and related processes, all of which work in tandem to achieve the goals of the organization (Hendrick and Kleiner). In the construction industry, smaller construction firms typically function as subcontractors. The vast majority of US construction firms are small businesses, meaning that 80% of general construction businesses employ 10 or fewer individuals (BLS, 2006). Typically, the organizational characteristics of small businesses differ from larger construction firms. Unlike larger construction firms, small construction businesses are less likely to implement safety training programs (Fielding and Piserchia, 1989; Eakins, 1992; Hollander and Lengermann, 1988; Holmes, 1995), and are also less likely to routinely and thoroughly communicate OSHA regulations to their workers (Williams, 1991; Rundmo, 1995). This latter deficiency may be due, in part, to the fact that small businesses with 10 or fewer workers receive a partial exemption to OSHA regulations. Consequently, these businesses may not be fully aware of their OSHA-related responsibilities. Moreover, these smaller firms operate in an extremely competitive business environments and lack the resources of larger construction firms (Peyton, 1996), which could also impact their ability to implement safety training programs. Additionally, smaller firms tend to be informal and do not always have explicit policies on safety and health. Another study by Smith-Jackson, Headen, Thomas, and Faulkner (2005), focusing on safety in small businesses, demonstrated the problems that accompany an informal work structure, which often has little state or federal oversight. This study elicited cultural critical incidents (Smith-Jackson and Leonard, 2003) from workers whose immediate supervisor was not of the same ethnicity. Most of the workers were African-American or Latino. An ethnic identity measure was used to assess workers' level of acculturation (or identification with the majority group), along with a psychometric scale to measure safety self-efficacy. Several interesting results were found. Ethnic identity was negatively correlated with safety self-efficacy; thus, the higher the individuals' identification with their own ethnic group, the lower their self-efficacy $[r_s (6) = -.77, p < .05]$. Content analysis of the interviews revealed several categories of cultural critical incidents in the occupational setting. Biased treatment was common, where supervisors assigned minority workers to more hazardous jobs or required minority workers to remain after the workday was over compared to White workers. It was commonly reported that employers did not compensate minority workers who were required to stay longer to close out the worksite. Inequities in access to training and access to PPE were also common occurrences. Additionally, workers reported losing interest in the job because supervisors overlooked or tolerated racist remarks or behaviors of other workers. These types of micro-aggressions (or slights by employers and targeted remarks or behaviors by workers) are not uncommon in the workplace, but more research should be done on the link between micro-aggressions and workplace safety and health outcomes.

TECHNOLOGICAL SUBSYSTEM

The technological subsystem comprises the tools, knowledge base, and technology required to complete the job done in a safe and efficient manner (Pasmore, 1988; Hendrick, 1991). When implementing new technology, possible effects on the social subsystem should be carefully considered before any changes are put into practice (Pasmore, 1988). The new technology invariably impacts both the organization and the individuals within it. The technological subsystem affects the social subsystem through the technical demands that help create the roles workers assume.

One major component of the technological subsystem is training. Although only a few research studies have investigated the problem of the disproportionately high injury and fatality rates among Latino construction workers, a number of reports have cited language barriers and the lack of training as additional factors (Brunette, 2004, 2005; Dong, Entzel, Men, Chowdhury, and Schneider, 2004; Goodrum and Dai, 2005; Nash, 2004; Pransky et al., 2002; Vasquez and Stalnaker, 2004). A community-based survey of Latino workers revealed both elevated rates of injury and lower rates of training compared to the non-Latino population (Pransky et al., 2002). Furthermore, the survey found that among the Latino immigrant workers who received training, only 25% received training in Spanish. A study conducted by Castillo, Davis, and Wegman (1999) concluded that Latinos were particularly susceptible to workplace hazards because of a lack of experience and training in the construction industry. The workers' limited on-the-job experience and training deficiencies made it difficult for them to recognize hazards and make proper judgments about risk taking on the job (Castillo et al., 1999; O'Connor, Loomis, Runyan, dal Santo, and Schulman, 2005). Additionally, most Latino workers were unaware of their legal rights as workers and lack the necessary confidence to speak up to supervisors about hazards in the workplace (NIOSH, 1999). This lack of safety training for Latino workers supports the social exchange theory that foreign workers receive fewer tangible and intangible outcomes from their employers (Ang et al., 2003; March and Simon, 1958).

Research by Smith-Jackson, Wogalter, and Quintela (2010) focused on Latino and Anglo (White) workers in crop production. Questionnaire ratings demonstrated significantly higher risk perception among Latino farmworkers, but a lower sense of confidence in protecting themselves. Lack of access to usable training and PPE was reported among Latino workers as one of the primary barriers to safety in crop production. It has been postulated that training non-Latino workers on safety would provide a sense of reciprocity, resulting in a safer workplace and higher perceived organizational support and distributive justice.

Ethnic and gender minority membership is related to belief systems and attitudes known to influence risk perception, health, safety climate, and self-protective behavior. By understanding the complex interplay between these factors and other factors as yet unidentified, we can design more effective interventions that are culturally centered and socially valid.

Besides safety and health beliefs, cultural differences may account for differences in communication patterns that are used to support work processes. We posit that cultural incompetence, indifference, and lack of awareness are intangibles that contribute to disparities in workplace injuries and fatalities. These intangibles (or sociocultural differences) account for disparities because of differences in work-related communication patterns between diverse or culturally incongruent workers that are not considered when designing training systems, risk communications, or work processes and protocols. Studies of cultural, behavioral, and cognitive patterns indicate that cultural markers exist that are elements of communication between similar or congruent cultural groups. These same markers can be misunderstood when exchanged between dissimilar or incongruent cultural groups (Baugh, 1983; Gudykunst, 1994; Larkey, 1996; Triandis, 1976). Communicative interactions in a workplace context of cultural incongruence have been shown to result in higher risks for hazard exposures.

Although the exploration of communicative interaction is relatively new in occupational safety and health, a few studies exist that have explored relationships to workplace safety and health. In an analysis of construction companies in Australia and Singapore, Loosemore and Lee (2002) found that construction companies tended to perceive cultural differences negatively, rather than perceive the differences as opportunities for new ways to interact. In addition, training programs in most construction companies surveyed were not provided in the indigenous languages of the workers. Finally, Loosemore and Lee suggested that the mismanagement of cultural diversity in workplace environments can contribute to lower quality of work life and problems for safety and health. Other studies by Mills (1972) and Lim and Alum (1995) showed a relationship between mismanagement of diversity and worker productivity, safety, and health in the construction industry. Future research in cultural ergonomics must examine interaction and risk mental models to identify cognitive and communicative differences between cultures that may help to understand the disparities in workplace fatalities and injuries.

Technologies are also artifacts, reflecting the preferences, lived experience, and histories of various nations and ethnic groups. In the workplace, pictorial symbols are used to communicate hazards and risks to workers, but rarely are cross-cultural differences considered in the design of such technologies as pictorial symbols. Instead, designers rely on "training" to ensure workers are aligned with a universal symbol system. Since access to effective training is not equitable, and since workers bring to the workplace mental models and implicit frames of reference that shape cognition, it is almost impossible to use training to overcome the potential risks introduced by a culturally biased pictorial symbol. As an example, hazard connotation research conducted by Smith-Jackson and Essuman-Johnson (2002) in Ghana, West Africa, revealed differences in symbol perceptions among factory workers, compared to the original intent of the symbols (Figure 6.2). Some of the differences could be classified as critical confusions—where the interpretation of the symbol was the exact opposite of the intent of the designer.

FIGURE 6.2   Skilled factory worker in Ghana. Workers comprehension of commonly used safety symbols was poor. Critical confusions also occurred.

The interpretations and hazard connotations of Ghanaian industry and trade workers were elicited using an open-elicitation procedure. Open elicitation gives no usage context for the risk communication component and relies solely on a direct conveyance of meaning and hazard connotation. Although most Ghanaians speak fluent English validity was upheld by using code-switching during the participant interviews using an interpreter/translator who code-switched (verbal and nonverbal) into Twi, Ga, Akan, or Fanti as appropriate. With the exception of the skull symbol, very few of the symbols were comprehended. For most of the symbols, the subsequent interpretations did not match the intended meaning of the symbols. Although all six symbols were designed to convey a significant degree of overall hazardousness, most of the responses did not convey any significant overall impressions of hazardousness.

Pictorial symbols are not the only biased technologies in the OSH setting. Colors used to connote hazards may also be biased and only meaningful to a subset of workers. In a study by Smith-Jackson and Wogalter (2000), comparisons of primary English speakers who were Caucasian/White and Latinos using independent group t-tests revealed significant differences between hazard ratings of certain symbols and colors. Thus, differences were demonstrated by the different hazard connotations assigned to colors between the groups. RED was equally rated among the two groups. However, orange and yellow were given significantly different ratings between the groups, with orange being rated as connoting more hazard than yellow, but this pattern was reversed among Caucasian/White workers. Some would suggest that workers can essentially adapt to these mismatches in perceptual-cultural connotations and colors used in the workplace. Training might assist workers to attach new meanings to new colors, but training may not be effective enough to alter the rapid, subconscious connotations triggered by specific colors. In other words, many of us in the United States may adapt to green stop signs, but the question is whether our braking speed will be as fast for a green stop sign (after years of exposure to them) compared to a red stop sign. Those few milliseconds of a difference in braking speed could mean the difference between life and death.

Studies of cultural difference in the workplace that impact OSH are much more advanced in other countries compared to the United States, especially Australia and the United Kingdom. Without a stronger research agenda for OSH that broadens the study of cultural differences in the workplace beyond narrowly studying disparities, we will not make progress in understanding the highly critical role of culture in all facets of OSH (e.g., safety climate, safety culture, micro-aggressions, personal protective technology fit and effectiveness, and training systems design). However, the research enterprise can continue to expand in this area without being totally constrained by a lack of funding at the federal level. One way researchers are collaborating to advance the field in the absence of adequate funding is by exploring the effectiveness of methods that can be used to conduct inclusive research in OSH.

CULTURAL ERGONOMICS METHODS IN OSH

Participatory ergonomics is an important method that can be used to reduce problems in a variety of OSH fields. Addressing social, psychological, and cultural needs of a given working population has been an important aspect of participatory ergonomics. Given the diversity of workplaces, time limitations, and financial constraints, it can be difficult to fully immerse oneself in a culture. Therefore, consulting with an expert is a way to benefit from the knowledge and experience of someone with prolonged and firsthand knowledge of the culture to be studied. If the individual is truly knowledgeable about the culture in question, this expert can sensitize the researcher, practitioner, or employer to facilitate the design of a culturally competent study, design, or intervention. Chapter 1 of this volume provides more coverage on issues and methodologies. In this chapter, we highlight a particularly effective method that has worked well in efforts to address cultural ergonomics in OSH settings, especially when comparing Latino workers to non-Latino workers.

Participatory ergonomics is a complementary method to help researchers, practitioners, and employers in OSH design culturally appropriate instruments, training, tools, and equipment.

Participatory ergonomics is defined as *the involvement of people in planning and controlling a significant amount of their own work and activities, with sufficient knowledge and power to influence both processes and outcomes to achieve desirable goals* (Wilson, 1995, p. 37). When designing culturally appropriate instruments, it is necessary to involve individuals from the cultural groups being targeted for the study. For the Latino population, it is recommended that participants come from different dialects of Spanish language used within the community and represent low and middle education levels or social classes.

Participation from individuals in the targeted group in the design of the instrument can provide the researcher insight into the following concepts of cultural adaptation:

- *Conceptual equivalence*: Do people attach the same meanings to terms and concepts (Stewart and Napoles-Springer, 2000)?

- *Cultural equivalence*: Are the cultural norms, beliefs, values, and expectations the same for different populations (Stewart and Napoles-Springer, 2000)?
- *Linguistic equivalence*: Do the words and grammar have similar meanings across different cultures and languages (Geisinger, 1994; Sperber, Devellis, and Boehlecke, 1994)?
- *Metric equivalence*: Do the numbers, counting, or scaling techniques mean the same thing or anything at all (Geisinger, 1994; Sperber et al., 1994)?

KEY PRINCIPLES FOR APPLYING CULTURAL ERGONOMICS TO OSH RESEARCH AND PRACTICE

Implementing the knowledge gained about cultural adaptation from the participatory approach will improve the cultural appropriateness of instruments, training, tools, and equipment. Below are 10 recommendations that can be used to ensure an inclusive environment for workplace safety:

1. **Have a "bilingual champion" on your research or design team:** In a bilingual environment, if the primary researcher, practitioner, or employer is not bilingual, it is imperative to have a bilingual individual on the team. Champions are in the best position to encourage a targeted group of workers (e.g., women, targeted group) to participate in the design process. The purpose of the champion is to build trust in the Latino community and to serve as the voice of the primary researcher, practitioner, or employer. Additionally, the research champion should know all the details of the project. These details include, but are not limited to, demographics of the workers (e.g., education level, socioeconomic status), the purpose of the project, benefits to the employer or workers, criteria and procedures for workers, and any instruments that will be used in the project.

2. **Build trust and relationships with prospective research participants:** To gain access to any targeted community that is to be the subject of a project or research study—it is necessary for the researcher, practitioner, or employer to build trust within that community by building relationships with potential workers. Building trust and building relationships are essential to the successful execution of any project. For the Latino and African-American community, building trust in people is an especially important core value (Vásquez and Stalnaker, 2004) that can facilitate greater access to workers. This process of building trust should begin at least 6 months prior to the onset of a project. To help build trust and subsequent relationships in the community, the researcher, practitioner, or employer and project champion should try to immerse themselves in the targeted community. Although physical immersion into a culture and exposure to daily lifestyle is preferred, there are other opportunities for immersion that require less time and financial resources. These activities include attending interest group meetings, dining at restaurants where the targeted workers dine, and taking part in cultural events. These strategies will allow the community to become familiar with the project, thereby enabling the researcher,

practitioner, and employer to build a network of individuals and meet key advocates in the community. Becoming known in the targeted community and networking with individuals prior to the project helps provide easier access to workers.

3. **Maintain short-term and long-term contacts with workers:** When building relationships with the targeted community, it is important to maintain short-term (e.g., 1–3 months) and long-term (6–12 months) contact with workers. Short-term contact is necessary once participants have agreed to assist with the project. For example, a month or more before the project is to be conducted, it is important to call or visit workers roughly every two weeks to engage with them, remind them about their commitment to the project, and to recruit other workers (snowballing) interested in participating in the project. Furthermore, contact with the workers one to two days prior to the project is needed to clarify the logistics of the project. In the case of construction work, unpredictable weather and deadlines can alter daily activities, and transportation is often an issue. Therefore, calling mobile phones five to eight hours prior to the project study is essential to ensure workers will be available at the appointed time and place.

   Long-term contact is also essential when a project requires the manipulation of data longitudinally. For example, some projects require multiple data collection sessions at various intervals; therefore, maintaining long-term contact is critical for retaining workers for the existing or future projects because they may be highly transient. Keeping in touch with workers demonstrates a genuine interest in them, making them more willing to assist the researcher, practitioner, or employer in finding other workers for future projects. Ways to maintain long-term contact with these workers include calling them occasionally after completing the project, visiting the workers' worksite, sending cards on appropriate holidays, engaging them as members or partners in your organization (e.g., Center for Innovation in Construction Safety and Health at Virginia Institute of Technology), and remaining active in their respective community.

4. **Know the language of the targeted group:** Although the project team may consist of a individual speaking the primary language of the target group, it is critical for all team members to have some familiarity with the language. It is not necessary to be fluent in the language, but it is essential to know key phrases (e.g., *hello, good-bye, my name is, I am from, how are you, thank you, you are welcome, I speak very little Spanish*). Team members should be willing to try to converse in at least a minimal way in the worker's native language with them. Trying to communicate in Spanish or knowing common phrases in Spanish demonstrates the researcher's willingness to learn Spanish and communicate in Spanish. In turn, individuals are likely to be more receptive to working with the researchers.

5. **Gain support from authority figures:** When dealing with the construction industry, whether it is residential or commercial, it is important to gain the support of the person in charge of the construction project (superintendent, project manager, and/or general contractor). To gain support, the

researcher, practitioner, or employer should come prepared to share the important details of the project with the appropriate authority figure. These details include the purpose of the project, estimated length of time, number of workers needed, when the project will take place (including time of day), what is required of the workers, and the type of compensation the workers will receive. Time is essential in this industry so the researcher, practitioner, or employer should try to explain everything succinctly in just 5–10 minutes. After gaining support from an authority figure, the researcher, practitioner, or employer should use this figure of authority to inform the subcontractors and workers on his or her site about the project. If potential workers are encouraged by their supervisor or an authority figure to participate in the project, it is easier to recruit dependable workers. To gain support and to recruit workers for this study, the researcher, practitioner, or employer and the project champion should visit different construction sites and make contact with the general contractors of construction projects and supervisors of crews.

6. **Explain relevance to targeted community:** A key factor in recruiting workers from a targeted community is to inform the workers about the importance of the project and how the project could benefit him or her as a worker in the targeted community. As long as the individual is valued, treated with respect, and can understand how they are contributing to the project, they are likely to be a loyal resource to assist with the project. For targeted ethnic groups, such as Latinos, these individuals usually enjoy taking part in projects because they feel they are being accepted in the United States and that non-Latino individuals care about their well-being. This type of treatment is, unfortunately, atypical for Latino construction workers because of the discrimination and disrespect they often face in the United States (Vásquez and Stalnaker, 2004). To explain why the project is relevant to the community, the researcher, practitioner, or employer and project champion should always present important statistics demonstrating the importance of this project before discussing the worker's participation. After discussing the statistics and why the project is focused on a targeted group, it should not be difficult to convey the importance of the project to the targeted group of workers.

7. **Understand the nature of the work environment:** When it comes to conducting research in the construction and crop production environments, it can be very frustrating because of the relative lack of formality in these industry sectors. For example, start/finish and break times can be highly variable from job site to job site. The researcher should be flexible and understand that a variety of unanticipated circumstances can postpone or even derail data collection. These circumstances include schedule changes, weather, fatigue, and high employee turnover. It is not unusual for a construction or agricultural/farm project to fall behind schedule, making it difficult for a worker to find time to contribute to the study. Conversely, if a project finishes ahead of schedule, the worker may leave early and be unavailable at the appointed time. Under certain weather conditions,

construction and agricultural workers may not come to work at all or may leave work early. Moreover, since these are manual labor-intensive fields, at the end of the workday, a participant may be too physically exhausted to contribute to the study, requiring the researcher to postpone and reschedule. To understand the nature of the work environment, prior to the study, researchers should interview and/or shadow construction and agricultural workers and supervisors to learn more about the demands and structure of the topic he or she is researching.

8. **Conduct the project during lunch or after the workday has ended:** If the project involves more than visual observation, the best time to conduct the project is during the lunch hour. If possible, the researcher, practitioner, or employer should provide lunch for the workers. Prior to providing lunch, the researcher, practitioner, or employer would want to know what is the most common type of food the targeted ethnic group of workers consume. Providing lunch is a nice gesture to show appreciation, and a free lunch often motivates workers to participate. Alternatively, if a project cannot be conducted during the lunch hour, the next best time to conduct the study is immediately after work.

9. **Take referrals from current workers:** Referrals are an excellent recruitment tool. After successfully finding workers for the project, always ask the workers if they know other workers who may be interested in participating in the project. Most workers will know others that may want to share their experiences. If they had a good experience during the research interaction, they may be more likely to be willing to tell other workers about the study. As an alternative to taking referrals, the researcher, practitioner, or employer can ask the workers to post flyers advertising the project in common areas where the targeted group convenes.

10. **Expect the unexpected:** Always expect the unexpected. When targeting specific groups for feedback and user experience, a researcher, practitioner, or employer may experience no-shows, or perhaps even more people than originally scheduled for the project. Moreover, a researcher, practitioner, or employer may experience rejection. No matter how persuasive the researcher, practitioner, or employer and/or project champion is, or how important the project is likely to be to the targeted community, there will be individuals who will want no part of the project. The researcher, practitioner, or employer should not become discouraged, but accept the fact that there are likely to be unpredictable periods during the project.

One final caveat relevant to field research in OSH with vulnerable populations is the necessity of being completely aware of the politics of the workplace. Employers, owners (farm, construction), and supervisors are not always accepting of research, especially when it involves interactions with people they employ. There are several reasons for this resistance. Field research can be intrusive, especially activities such as shadowing. It is important to seek the approval of employers or relevant parties before beginning. Researchers should be aware that their presence may trigger work stoppages or slow-downs, which, for small companies, could be costly. Additionally, the presence of researchers at work sites can introduce additional hazards, so the design of studies utilizing observational methods should be considered carefully. Employers are often concerned about the likelihood the research will uncover violations, especially workplace safety violations. Researchers should consult with their human subjects ethics offices to acquire advice on the levels of confidentiality covered under the ethics regulations and guidelines.

Politics will also play a role due to the power dynamics of the workplace. Workers, especially ethnic minorities and low-income White workers, are vulnerable to coercion by employers. Thus, it is important that researchers ensure they acquire the informed consent of workers, and should not only acquire the consent of employers. Workers also worry about retaliation, in spite of confidentiality requirements associated with research ethics. One way to make workers feel more comfortable is to collect data off-site to the greatest extent possible, and to arrange meetings with workers. Meeting in community locations and other familiar environments will make a difference in lowering the anxiety of workers and acquiring detailed and rich data. Similarly, public events such as festivals or flea markets usually yield successful recruitment, especially when compensation for time is provided. Monetary compensation is very important, although researchers should be cognizant of the ethics of setting compensation levels so they are not excessive but also not so minimal as to be nonmotivating.

REFERENCES

Abrams, A. (2006). Gender-specific hazards: Global applications for occupational safety and health research and policy development. ASSE Professional Development and Exposition, June 11–14, 2006, Seattle, Washington. American Society of Safety Engineers.

Anderson, J. T. L., Hunting, K. L., and Welch, L. S. (2000). Injury and employment patterns among Hispanic construction workers. *Journal of Occupational and Environmental Medicine,* 42, 176–186.

Ang, S., Van Dyne, L., and Begley, T. M. (2003). The employment relationships of foreign workers versus local employees: A field study of organizational justice, job satisfaction, and OCB. *Journal of Organizational Behavior,* 24, 561–583.

Baugh, J. (1983). *Black street speech: The history, structure, and survival.* Austin: University of Texas Press.

Brunette, M. (2005). Development of educational and training materials on safety and health: Targeting Hispanic workers in the construction industry. *Journal of Family and Community Health,* 28, 253–266.

Brunette, M. (2004). Construction safety research in the United States: Targeting the Hispanic workforce. *Injury Prevention,* 10, 244–248.

Bureau of Labor Statistics (BLS) (2006). *Construction in Occupational Handbook.* Washington, D.C. Bureau of Labor Statistics (BLS). Retrieved July 1, 2007, from http://www.bls.gov/news.release/cfoi.nr0.htm

Castillo, D. N., Davis, L., and Wegman, D. H. (1999). Young workers. *Occupational Medicine,* 14, 519–536.

Diekman, A., and Eagly, A. (2000). Stereotypes as dynamic constructs: Women and men of the past, present, and future. *Personality and Social Psychology Bulletin,* 26, 1171–1188.

Dong, X., Entzel, P., Men, Y., Chowdhury, R., and Schneider, S. (2004). Effects of safety and health training on work-related injury among construction laborers. *Journal of Occupational Environmental Medicine,* 46, 1222–1228.

Dong, X., and Platner, J. W. (2004). Occupational fatalities of Hispanic construction workers from 1992–2000. *American Journal of Industrial Medicine, 45*–54.

Eakins, J. (1992). Pictorial information systems—prospects and problems. *Proeedings of the 14th British Computer Society Information Retrieval Specialist Group Research Colloquium on Information Retrieval.* Lancaster University, UK, pp. 102–123.

Ennis, S., Rios-Vargas, M., and Albert, N. (May 2011). *The Hispanic Population: 2010. U.S. Census Briefs.* Retrieved from http://www.census.gov/prod/cen2010/briefs/c2010br-04.pdf, December 15, 2011.

Fabrego, V., and Starkey, S. (2001). Fatal occupational injuries among Hispanic construction workers of Texas, 1997–1999. *Human Ecological Risk analysis, 7,* 1869–1883.

Fielding, J., and Piserchia, P. (1989). Frequency of worksite health promotion activities. *American Journal of Public Health, 79,* 16–20.

Geisinger, K. (1994). Cross-cultural normative assessment: Translation and adaptation issues influencing the normative interpretation of assessment instruments. *Psychological Assessment, 6,* 304–312.

Goodrum, P., and Dai, J. 2005. Differences in occupational injuries, illnesses, and fatalities among Hispanic and non-Hispanic construction workers. *Journal of Construction Engineering and Management* 131(9), 1021–1028.

Gudykunst, W. B. (1994). *Bridging differences: Effective intergroup communication, 2nd ed.* Thousand Oaks, CA: Sage.

Hendrick, H. W. (1991). Ergonomics in organizational design and management. *Ergonomics, 34,* 753–756.

Hendrick, H. W., and Kleiner, B. M. (2001). *Macroergonomics: An Introduction to Work System Design.* Santa Monica, CA: Human Factors and Ergonomics Society.

Hendrick, H. W., and Kleiner, B. M. (2002). *Macroergonomics: Theory, Methods and applications.* Mahwah, NJ: Lawrence Erlbaum Associates.

Holt, C., and Ellis, J. (1998). Assessing the current validity of the Bem Sex-Role Inventory. *Sex Roles,39,* 929–941.

Jackson, S., and Loomis, D. (2002). Fatal occupational injuries in the North Carolina construction industry, 1978–1994. *Applied Occupational and Enviornmental Hygiene, 17,* 27–33.

Larkey, L. K. (1996). Toward a theory of communicative interactions in culturally diverse workgroups. *Academy of Management Review, 21,* 463–491.

Lim, E. C., and Alum, J. (1995). Construction productivity: Issues encountered by contractors in Singapore. *International Journal of Project Management, 13,* 51–58.

Loh, K., and Richardson, S. (2004). Foreign-born workers: Trends in fatal occupational injuries, 1996–2001. *Monthly Labor Review, 42*–53.

Loosemore, M., and Lee, P. (2002). Communication problems with ethnic minorities in the construction industry. *International Journal of Project Management, 20,* 517–524.

March, J. G., and Simon, H. A. 1958. *Organizations.* New York: John Wiley & Sons.

Mills, D. Q. (1972). *Industrial Relations and Manpower in Construction.* London: MIT Press.

Muntaner, C., and Schoenbach, C. (1994). Psychosocial work environment and health in U.S. metropolitan areas: A test of the demand-control and demand-control-support models. *International Journal of Health Services, 24,* 337–353.

Nash, J. 2004. Construction safety: Best practices in training Hispanic workers. *Occupational Hazards* 66(2), 35–37.

National Institute for Occupational Safety and Health (NIOSH) (1999). *Promoting Safe Work for Young Workers: A Community Based Approach.* Publication number 99–141.

O'Connor, T., Loomis, D., Runyan, C., dal Santo, J. A., and Schulman, M. (2005). Adequacy of health and safety training among young Latino construction workers. *Journal of Occupational and Environmental Medicine, 47,* 272–277.

Pasmore, W. (1988). *Designing Effective Organizations.* New York: John Wiley & Sons.

Peyton, G. (1996). Small business versus risk. *Australian Safety News*, August, 16–17.

Pransky, G., Moshenberg, D., Bejamin, K., Portillo, S., Thackery, J. L., and Hill-Fotouhi (2002). Occupational risks and injuries in non-agricultural immigrant Latino workers. *American Journal of Industrial Medicine,* 117–123.

Prentice, D. A., and Carranza, E. (2004). Sustaining cultural beliefs in the face of their violation: The case of gender stereotypes. In M. Schaller and C. Crandall (Eds.), *The Psychological Foundations of Culture*, pp. 259–280. Mahwah, NJ: Erlbaum.

Roanoke Times (Jan. 17, 2004). Bricklayer won't let her dream crumble on her. Edition Metro Dateline, Virginia Beach, VA, page B:8.

Rudman, L., and Glick, P. (1999). Feminized management and backlash toward agentic women: The hidden costs to women of a kinder, gentler image of middle managers. *Journal of Personality and Social Psychology,* 77, 1004–1010.

Rundmo, T. (1995). Perceived risk, safety status, and job stress among injured and noninjured employees on offshore petroleum installations. *Journal of Safety Research,* 26, 87–97.

Sasao, T., and Sue, S. (1993). Toward a culturally anchored ecological framework of research in ethnic-cultural communities. In E. Seidman, D. Hughes, and N. Williams (Eds.), *Special Issue: Culturally Anchored Methodology. American Journal of Community Psychology,* 21, 705–728.

Smith-Jackson, T. L., and Wogalter, M. S. (2000). Users' hazard perceptions of warning components: An examination of colors and symbols. *Proceedings of the 14th Triennial Conference of the IEA/HFES,* 6, pp. 55–58.

Smith-Jackson, T. L., and Essuman-Johnson, A. (2002). Cultural ergonomics in Ghana, West Africa: A descriptive survey of industry and trade workers' interpretations of safety symbols. *International Journal of Occupational Safety and Ergonomics,* 8, 37–50.

Smith-Jackson, T. L., Headen, E., Thomas, C., and Faulkner, B. (2005). Cultural critical incidents in hazardous occupations: A preliminary exploration. In P. Carayon, M. Robertson, B. Kleiner, and P. L. T. Hoonakker (Eds.), *Human Factors in Organizational Design and Management–VIII,* pp. 479–484, Santa Monica, CA: IEA Press.

Smith-Jackson, T. L., Wogalter, M. S., and Quintela, Y. (2010). Safety climate and pesticide risk communication disparities in crop production by ethnicity. *Human Factors and Ergonomics in Manufacturing,* 20, 511–525.

Sperber, A., Devellis, R., and Boehlecke, B. (1994). Cross-cultural translation: Methodology and validation. *Journal of Cross-Cultural Psychology,* 25, 501–524.

Stewart, A., and Napoles-Springer, A. (2000). HRQL assessment in diverse population groups in US. *Medical Care,* 38, 102–124.

Tepper, B., Brown, S., and Hunt, M. (1993). Strength of subordinates' upward influence tactics and gender congruency effects. *Journal of Applied Social Psychology,* 23, 1903–1919.

Triandis, H. C. (1976). Approaches toward minimizing translation. In R. Breslin (Ed.), *Translation: Applications and Research*, pp. 229–243. New York: Wiley/Halstead.

Vázquez, R., and C. Stalnaker (2004). Latino workers in the construction industry: Overcoming the language barrier improves safety. *Professional Safety,* 49, 24–27.

Williams P. (1991). Planning factors contributing to ongoing health promotion programs, *Journal of Occupational Health and Safety—Australia and New Zealand,* 7, 489–494.

Wilson, J. R. (1995). Ergonomics and participation. In J. R. Wilson and E. N. Corlett (Eds.), *Evaluation of Human Work*, pp. 1071–1097. London: Taylor & Francis.

7 Military Systems

William Lee

CONTENTS

INTRODUCTION[*]

The Department of Defense (DoD), formerly known as the War Department, was created by the National Security Act of 1947 (P.L. 80-235, 61 Stat 496) in July of that year (National Security Act of 1947 [1947]). Signed by President Harry Truman, the act essentially merged the War Department and the Navy Department into what was initially known as the National Military Establishment, headed by the Secretary of Defense. Simultaneously, this act also established an independent US Air Force (USAF) as a coequal executive department to the army and navy. Two years later, in 1949, the National Military Establishment was renamed the Department of Defense. For further background information on the establishment of the DoD, refer to the document, "Toward Independence: The Emergence of the U.S. Air Force 1945–1947" (Wolk, 1996).

In addition to housing the US Army, Navy, Marine Corps, and Air Force, a handful of other major agencies are also under the purview of the DoD. Importantly, the DoD plays a vital role in national security in the collection of intelligence through its Defense Intelligence Agency[†]. It also engages in the electronic collection of intelligence through the National Security Agency[‡] and satellite organizations like the National Geospatial-Intelligence Agency[§] and the National Reconnaissance Office[¶]. These various DoD arms make it the single largest agency of the US federal

---

[*] Approved for Public Release: 11-3720. Distribution Unlimited.
[†] http://www.dia.mil/.
[‡] http://www.nsa.gov/.
[§] https://www1.nga.mil/Pages/Default.aspx.
[¶] http://www.nro.gov/.

government with the largest number of employees. In fact, in 2008, its slice of the federal budget was 21%, even after excluding appropriations for the wars in Iraq and Afghanistan and veteran-related obligations. Thus, no other government agency approaches the DoD in terms of influence, responsibility, and global representation. It is widely reported, for example, that US troops are currently stationed in 177 countries (over 70% of the world's countries). With troops and personnel in so many foreign locations, the DoD has recognized the need for improved cultural and social knowledge within the armed services—and is taking steps to educate personnel in various ways. The problem, however, has been to access, coordinate, and disseminate cultural knowledge to the thousands of individuals whose effectiveness could be improved by enhanced cultural awareness.

While the goal of achieving greater cultural competency may seem as straightforward as learning the local language, the process is far more nuanced and should begin with a viable definition of "culture." Within the context of the DoD, however, the concept of culture is difficult to define. According to Veroff and Goldberg (1995), culture is defined as "a collectivity of people who share a common history, often live in a specific geographic region, speak the same or a closely related language, observe common rituals, beliefs, values, rules, and laws, and which can be distinctively identified according to cultural normative practices." Therefore, based on this definition and the challenges the DoD faces with respect to making its workforce more culturally competent, we operationalize culture as follows: *the inclusive knowledge that is required by the DoD to more effectively complete its domestic and overseas missions.*

IMPORTANCE OF CULTURE FOR TODAY'S WARFARE

Why has the concept of culture become so important to the DoD at the present time—especially in light of America's long history of engaging in foreign wars? Indeed, the United States was born from conflict with the British. To answer this question, it is important to briefly review lessons learned from previous and current conflicts. Col. Carolyn F. Kleiner, director of Strategic Studies and Research, Reserve Components and the Senior US Army Reserve Advisor to the Strategic Studies Institute at the US Army War College, provided one of the best summaries of the role of culture in previous global conflicts involving the United States (Kleiner, 2008). Traditionally, when the United States planned conflicts and engaged opponents in foreign wars, it was against forces of *similar* backgrounds—primarily European cultural backgrounds. After the Revolutionary War, the War of 1812 was also successfully waged against the British, followed by World War I (Germans), and then World War II in two theaters of battle, Europe and Japan. It could be argued that the growing importance of cultural differences became more apparent during the war in the Pacific when American soldiers were brutally challenged by unfamiliar Japanese military tactics. As US forces eventually discovered, Japanese soldiers adhered to the medieval code of *bushido*— a code of honor that can be closely compared to the medieval European notion of chivalry. This bushido mentality made them relentless opponents who were willing to fight to the death.

The importance of culture was also seen and experienced by officers and enlisted personnel during the Vietnam War (1964–1973). As Gen. Anthony C. Zinni* discussed before deploying to Vietnam, he attended the Military Assistant Training Advisors (MATA) course at the John F. Kennedy Special Warfare Center in Ft. Bragg, North Carolina. During that time, he attended language courses taught by native Vietnamese speakers. In addition to acquiring language fluency, interacting with his Vietnamese instructor also exposed him to important cultural information. Moreover, when not in formal training, Zinni mingled with local Vietnamese residents at their restaurants to further develop his knowledge of the history and cultural traditions of Vietnam. Even after he was deployed, he continued to interact with Vietnamese marines who enhanced his knowledge of the country and its people. Despite the losses of that war, the cultural lessons that Gen. Zinni gained from that experience shaped his subsequent interactions. Notably, over time he learned to develop an appreciation and understanding of non-Western cultures. Zinni credits his Vietnam experience with future successes in understanding the subtle political and cultural differences that have shaped and continue to shape conflicts in volatile areas, such as the Middle East.

Despite lessons learned in Japan and Vietnam about the importance of understanding the cultural traditions and beliefs of the adversary, it was the US invasion of Iraq that "served as a wake-up call to the military that adversary culture matters" (McFate, 2005). In short, with the US military's involvement in both low- and high-intensity warfare in Iraq and Afghanistan (and related humanitarian operations), culture has emerged as an important factor. Early on in the Iraq war, for example, experts soon realized that although the conflict was being fought brilliantly at the technological level, gains at the human level were inadequate at best. The war in Afghanistan—against an enemy who fights unconventionally—reinforced the need to understand motivation, intent, method, and culture.

Thus, the DoD is engaging in what is known as "culture-centric warfare" in which soldiers, analysts, and commanders are being made aware of local social norms and nuances to tackle the "human piece" of contemporary conflicts.

Basic Lessons from the Field

According to Michael Lewis (2010) in his article "The Army's Foundational Building Blocks of Soldier Skills," the US Army Training and Doctrine Command revised one of its fundamental tenets—the Warrior Tasks and Battle Drills—with the goal of understanding "the basics of the foreign culture, including religious factors, social intelligence, and cultural behaviors." Gen. David H. Petraeus, as commander of the NATO International Security Assistance Force and US Forces Afghanistan, once referred to knowledge of a foreign culture as a "force multiplier" (Petraeus, 2006). Specifically, according to Gen. Petraeus's observation in Iraq, "knowledge of the

---

* John A. Adams, '71 Center for Military History and Strategic Analysis, Cold War Oral History Project. Interview with Gen. Anthony C. Zinni by Cadet Shelby Sears, June 29, 2004, http://www.vmi .edu/uploadedFiles/Archives/Adams_Center/ZinniAC/ZinniAC_interview.pdf

cultural 'terrain' can be as important as, and sometimes even more important than, knowledge of the geographic terrain. This observation acknowledges that the people are, in many respects, the decisive terrain, and that we must study that terrain in the same way that we have always studied the geographic terrain." Crane et al. (2009) also argued that in stability operations, understanding economic forces in the local culture can result in important outcomes, such as a reduction in crime and insurgency momentum, increased goodwill toward occupying forces, the greater likelihood of lasting peace—and ultimately the timely withdrawal of troops. Something as obvious as understanding the local currency and what constitutes a living wage in the local economy, as well as training/hiring locals to fill available jobs, could have long-lasting positive consequences. Additionally, with trade being an important source of income for developing countries such as Iraq and Afghanistan, it is important for US military personnel to better understand traditional market cycles/practices, trading locations/routes, and commodities that drive the local economy so that efforts can be made to protect and grow it.

CULTURAL LESSONS LEARNED AND APPLIED

In addition to wartime operations, cultural awareness has also been vital for reconstruction efforts. According to Lt. Col. Michael B. Meyer (2009), the US military recognized the importance of cultural awareness in its attempt to help rebuild Japan after World War II. During the occupation of Japan, Gen. Douglas MacArthur and his staff operated two organizations in order to demilitarize and democratize the Japanese government. Of particular relevance to this chapter is the Public Opinion and Sociological Research Division (PO&SR), which employed social scientists who used applied anthropology (e.g., observation), interviews, and surveys to understand the unfamiliar Japanese history/customs so that leaders could make informed decisions. However, these PO&SR social scientists did not operate in a vacuum. Instead, the division worked with and trained its Japanese counterparts so successfully that the result was "a genuine feeling of camaraderie with its Japanese staffers."

In summary, Lt. Col. Mayer described four specific, culture-related lessons that were learned from the war in the Pacific and the ensuing Japanese occupation. *First, the training of culturally aware soldiers and related personnel must be reinforced by all available experiential and anecdotal information.* For example, the wartime experiences of American soldiers during various Pacific conflicts helped to shed light on the importance of "saving face" in Japan. Schools were established in the United States to train noncombat forces to prepare them for service in occupied Japan. The curriculum of these schools focused on how Japan's culture was different from that of the West—with particular emphasis on the importance of honor (i.e., bushido). *Second, during an occupation it is essential to develop an atmosphere of mutual respect and collaboration.* Notably, because of the genuine desire of American social scientists to help the Japanese people—and bolstered by their understanding of Japanese culture—they were better equipped to build rapport and trust. The result was that after the occupation the Japanese used what they learned from the Americans to continue to drive the democratization of their country. *Third, cultural intelligence was and remains the most important source of first hand information.*

With American social scientists working and living side-by-side with their Japanese counterparts, these scientists were able to provide analysts with actual firsthand, nonspeculative information. *Fourth, the U.S. military can collaborate in productive ways with academics and other nonmilitary experts to increase culture awareness.* Although the occupation of Japan was far from problem-free, it helped set the precedent for gathering information from nonmilitary sources, such as colleges, think tanks, etc. As a current example of this principle, the army instituted the Human Terrain System (HTS)[*], which employs social scientists and anthropologists who are deployed to field locations to serve as advisors for military leaders.

CURRENT CULTURE LANDSCAPE

Today's "Global War on Terror" has brought culture awareness to the forefront, making this term more than just a buzzword, but rather a mandate for improving military efforts at the tactical and operational level. As such, the DoD and its various service branches have all developed training programs for enhancing cultural awareness. Take, for example, the Air University Culture and Language Center[†] at Maxwell Air Base in Montgomery, Alabama (Air Force), the Advanced Operational Cultural Learning Center[‡] in Quantico, Virginia (Marine Corps), the Training and Doctrine Command Culture Center at Fort Huachuca, Arizona (Army), and the Defense Language Institute Foreign Language Center[§] in Monterey, California, as evidence that the DoD is taking this issue very seriously.

From the perspective of the Special Operations community, there is an effectiveness ceiling that goes beyond "culture awareness." According to Zahn and Lacey (2007) in their treatise "Building a Virtual Cultural Intelligence Community," these Special Operations soldiers often need to "comprehensively understand the socioeconomic, historical, and cultural landscape in which social and political movements—to include terrorist groups—live." Lacking a true comprehension of "the complexity and richness of the values and concerns of the people" could put missions at risk for failure. Zahn and Lacey referred to this kind of culture savvy as "culture intelligence" and differentiated it from culture awareness, which they defined as knowledge of a culture gained simply from information and facts (i.e., "what"). Conversely, culture intelligence is the result of an analytical cognitive process that would enable individuals to grasp the intent (i.e., "why") of their opponents.

One important and somewhat recent effort instituted by the DoD to enhance its cultural awareness impact is the push to expand the role of women in the armed forces. According to a five-part series (Norris, 2007) broadcast by National Public Radio, until fairly recently women have been serving in the military in largely supporting roles. However, as the scope of the Iraq and Afghanistan conflicts continued to expand, women have had more opportunities to serve in combat roles— although this expansion has proved controversial. In her study "Women Training

---

[*] http://humanterrainsystem.army.mil/.
[†] http://www.culture.af.mil/.
[‡] http://www.tecom.usmc.mil/caocl/.
[§] http://www.dliflc.edu/index.html.

Iraqi Soldiers and Police in Iraq: The War with Three Fronts," Maj. Michelle Stringer described her yearlong deployment in Iraq and the "three-front war" that women face: battling the resentment and prejudice of American military men, fighting the enemy, and dealing knowledgably with the markedly different culture for women that female service members face in Iraq (Stringer, 2008). As an experienced security officer, Stringer and other female officers schooled Iraqi male soldiers on issues related to physical security. As she discussed, language, religion, and Islamic beliefs continued to form barriers to effective outcomes—although, interestingly, they seemed to be the same barriers experienced by their male peers. Further, she also argued that existing cultural training is not adequate or effective. In one instance, for example, a female service member was observed to greet two tribal leaders with an appropriate culture-specific custom. However, the author later learned that the training this woman received prior to being deployed had not educated her with respect to the origin or implications of this custom—knowledge that she could have used in other situations. It took a local interpreter's "on-the-job" guidance to provide this information.

Although the importance of culture training has been stressed by the military and their senior leaders, skepticism still exists. From a systems perspective, Marine Corps Gen. James Mattis, head of US Joint Forces Command from 2007 to 2010, asserted that equipping soldiers with cultural knowledge cannot be accomplished merely through training and education—repeated deployments are essential for building cultural understanding. The problem, according to Mattis, is that "the system [personnel, promotions] does not reward cultural skills" (Erwin, 2009).

Hence, researchers have come to realize that they needed to go back and understand what was at stake and how to motivate our service members in the area of culture. In a comprehensive survey conducted by Szalma, Hancock, and McDonald (2009), more than 50,000 service men and women from every service branch (US Air Force, Army, Navy, and Marines) were polled on the "challenge of culture." The authors reported two interesting findings with respect to culture training. First, the respondents' attitudes toward the importance of culture for completing missions differed according to service branch and rank. Notably, the authors found that Air Force personnel tended to perceive the importance of culture more positively than personnel associated with the other four branches. Further, the authors found that junior enlisted personnel (i.e., ranks E1–E6) were more skeptical about the importance of culture compared to higher-ranked personnel. Second, recent prior experiences gained through deployment affected a respondent's attitude toward culture importance. For example, the authors indicated that currently or recently deployed Marines and enlisted Air Force personnel tended to perceive the importance of culture more negatively than personnel who had not been deployed lately. Although this finding conflicts with Mattis's assertion that "repeated deployments are essential for building cultural understanding," such negative attitudes do, in fact, reinforce the expectation model of culture awareness discussed by Wunderle (2006). Wunderle asserted that in the absence of significant exposure to the foreign culture, soldiers would likely experience negative emotions from their deployment due to confusion, frustration, and anguish. Unfortunately, even when soldiers are deployed for longer periods or for repeated tours of duty, their added culture experience does not necessarily contribute to a greater appreciation for culture awareness—due, in part, to the steep learning curve for cultural training, particularly if language fluency is taken into account.

In summary, these conflicting reports about the importance of culture—and how that varies by service branch, rank level, and on-the-ground experience—reinforce the complexity of nurturing cultural awareness among DoD personnel.

CURRENT APPROACHES IN CULTURAL TRAINING

As noted earlier, the need to train for cultural capability is not new. However, current military engagements against unfamiliar opponents using "nontraditional" tactics have highlighted the need to understand cultures and societies that are dramatically different from our own. A variety of approaches have been used to meet that goal. For example, the US Army's National Training Center at Ft. Irwin, California, uses mock Iraqi villages deep in the Mojave Desert to train soldiers before their actual deployment (Filkins and Burns, 2006). These villages are staffed with authentic actors (e.g., Arab Americans) to expose soldiers to the "blood and tears" of operating in a foreign environment. While such realistic training settings have helped to familiarize military personnel with some of the situations they are likely to encounter, more innovative approaches are also being tested.

Computer simulation has emerged as a "virtual cultural training ground" for military personnel—although the jury is still out with respect to the effectiveness of these approaches. As an example of this technique, Gratch and Marsella (2001) described the use of "virtual humans" (anthropomorphic computer characters [ACCs]) within a prototype system for teaching decision-making capabilities in volatile situations. The prototype system, known as the Mission Rehearsal Exercise, was designed to provide an immersive learning environment where trainees could experience the sights, sounds, and circumstances they would encounter in real-world scenarios while performing mission-oriented training. This highly sophisticated system used dozens of prescripted ACCs to play characters in a military peacekeeping exercise. Specifically, the ACCs were designed to look like Bosnian civilians and soldiers from a peacekeeping force. The system was intended to educate trainees on what to do and what not to do in a potentially volatile situation.

A similar system called the Virtual Environment Cultural Training for Operation Readiness (VECTOR) was created by McCollum et al. (2004) as a training platform for teaching soldiers culturally related interaction skills. Specifically, ACCs within VECTOR would express emotions bounded by preprogrammed social norms. This allowed a trainee soldier of VECTOR to become more familiar with the likely emotional responses of a particular culture. Further, Alelo, Inc.'s Tactical Language and Culture Training Systems employ gaming concepts and ACCs to teach languages and subtle nonverbal behaviors interactively in a simulated environment.

Language instruction—traditionally a mainstay of culture training—is also using advanced technologies involving simulation and gaming. The Tactical Language Training System (Johnson et al., 2004) uses computer-animated characters, speech-recognition technology, and real-life scenarios to allow trainees to practice communicating and interacting with virtual locals through verbal and nonverbal behaviors. In addition, trainees also receive feedback as a form of scaffolding from a virtual coach. Moreover, the Defense Language Institute Foreign Language Center in Monterey, California, educates DoD personnel in their eight separate language schools.* Their facilities can accommodate approximately 3,500 soldiers, marines, sailors, and airmen, as well as select DoD employees. To attend the center, one must be a member of the armed forces or be sponsored by a government agency. Students can learn languages in a variety of ways—from total immersion programs with native language speakers to use of their new Headstart program, which consists of an interactive 80-hour, self-paced DVD that teaches basic language, culture, and limited reading and writing skills. The avatar characters used in this product are designed to function along the lines of today's interactive computer games. Headstart is currently available in Iraqi Arabic and in Dari and Pashto (languages spoken in Afghanistan).

SERVICE BRANCHES' TRAINING FOR CROSS-CULTURAL COMPETENCE

In a review entitled "Language and Culture Training: Separate Paths?" (Watson, 2010), the author noted that because the various service branches have different definitions of what constitutes "intercultural effectiveness," each has created its own center for dealing with the demands of educating personnel for cross-cultural competence. However, their methods do not always agree. Specifically, while language competence is considered to be an important skill for some services, other branches minimize the importance of language fluency, which detractors define as "perishable and time-intensive to attain and sustain." Instead, they focus on cross-cultural competency, which "represents knowledge that is more durable and more easily attainable." Complicating this issue of cross-cultural competency is that there are differing perspectives even within a single service branch. For example, although the US Army's TRADOC Culture Center focuses on teaching cultural competence with a higher priority over developing language skills, the army's primary training ground—the United States Military Academy (USMA) at West Point (where Watson is employed)—recently created a Center for Languages, Cultures, and Regional Studies to encourage a boarder approach for enhancing cross-cultural competence. Notably, administrators of the West Point center believe in the education of "language, culture, and the knowledge of regional dynamics as vitally interrelated and equally important aspects of intercultural effectiveness." Although language facility is the cornerstone of the center, it does define cross-cultural competence as "the capacity to generate perceptions and to adapt behavior to cultural context"—traits not necessarily wedded to language fluency.

Similarly, other researchers agree that language training may not be the best avenue for reaching the goal of cross-cultural competence. Notably, Abrahams (2007) suggested that "emotional intelligence" (EI) is a vital interpersonal skill for enhancing cultural competency. EI is defined as the self-perceived ability to identify, assess, and manage the emotions of one's self, of others, and of groups. In other words, by recognizing and understanding their own emotion—as well as that of others—military leaders would be more adept at handling a variety of both social and combat

---

* http://www.dliflc.edu/schools.html.

situations through better communication and understanding. The Special Forces (SF) community within the US Army is, in fact, one of the communities that is leveraging emotional intelligence for selecting and training their members. Because SF personnel frequently work with members of local communities and governmental agencies, these soldiers need to possess broader interpersonal skills—including EI. Of particular importance to the concept of emotional intelligence is the idea of self-awareness. In short, SF soldiers with a high degree of self-awareness would be able to better perceive both verbal and nonverbal cues, which they could then use to adapt their behaviors for important tasks such as negotiations and collaborations with local personnel.

Two service branches that hold similar views about the importance of culture are the US Marine Corps (USMC) and the US Navy (USN)—although they differ in their views of the important of language training. The USMC established the Center for Advanced Operational Culture Learning, which supports a pragmatic view of culture. In essence, the USMC determined that culture training should only focus on operationally relevant concepts. In other words, any factors outside the USMC-defined areas of culture focus—including physical environment, economy, social structure, political structure, and belief systems—would not be considered as operationally relevant. It should also be noted that according to Watson (2010), language training was also omitted in the USMC's center definition of culture training. The USN, however, places a higher emphasis on foreign language facility. Under the Navy Language Skills, Regional Expertise and Cultural Awareness (LREC) Strategy (U.S. Navy Language Skills, Regional Expertise and Cultural Awareness Strategy, 2008), the USN takes a pragmatic view of language development. According to the strategy guide, the USN specifically leverages the concept of culture training with the Defense Language Transformation Roadmap. Through this roadmap, the USN aims to create a pool of career language professionals, sailors with working-level language skills, and reservist specialists with foreign language skills for emerging contingencies.

The USAF also places a high value on culture training, as evidenced by the establishment of their Air Force Culture and Language Center, which emerged as a result of the USAF-sponsored Cross Cultural Competence (3C) Support Study. Based on study outcomes, the USAF defined 3C as "the ability to quickly and accurately comprehend, then appropriately and effectively act, to achieve the desired effect in a culturally complex environment—without necessarily having prior exposure to a particular group, region or language." To achieve such competence, the USAF envisions that all airmen should possess a combination of *general* cultural (e.g., shared meaning), regional (e.g., state, geographical, and organizational), and language (e.g., communication through verbal and nonverbal) capabilities. However, for a variety of specific mission requirements, the USAF further asserts that airmen would need *specialized* knowledge of culture, region, or language.

Interestingly, the USAF model has been adopted by Canadian military personnel. Selmeski (2007), in his comprehensive discussion of "Military Cross-Cultural Competence: Core Concepts and Individual Development," described how the Canadian military has adopted his "Professional Development Framework" (PDF), which emerged from his study of the USAF 3C model. It should be noted, however, that the PDF does not focus on some formalized study of culture. Rather, the PDF

TABLE 7.1

Best Practices for Training Cultural Awareness

Experience	Encourage military personnel to live off base, travel, participate in diverse communities
	Recruit more minorities, bi-, multi-, and transcultural individuals
	Attract more members with foreign experience (e.g., former exchange students) and varied educational/professional backgrounds
	Provide military personnel stationed overseas with learning opportunities involving language, culture, and geography
Education	Place greater emphasis on the liberal arts, especially anthropology, religion, sociology, psychology, geography, and philosophy
	Expose personnel to culture concepts through tailored anthropology courses
	Teach basic anthropological methods to help individuals gather information on their own

Source: http://www.culture.af.edu/PDF/SelmeskiMay2007.pdf.

recognizes five interrelated capacities—expertise (technical/strategic knowledge), cognitive (analytic/creativity ability), social (communication, interpersonal skills), change (self/group, learning organization), and professional ideology (moral reasoning, credibility and impact). By applying the PDF model through training and education, the Canadian forces believe that their service members will be exposed to these various capability stages throughout their careers. As a result, the author asserted that when stressing and combining the five capacities in different situations, emerging competencies could then be developed. Yet, as the author admitted, the PDF does not explicitly address the need for culture competence. Hence, the author proposed that the PDF could, in fact, be combined with the USAF 3C model to help build three levels of professional development. The first is Initial/Novice. In this level, cadets and new officers would be encouraged to develop self-awareness and basic knowledge of culture concepts. The second level is Intermediate. In this level, intermediate-level soldiers (e.g., lieutenants, captains, squad leaders) would develop key cultural knowledge (e.g., religion), collaborate with others, realize others' values, and adapt their behaviors contextually. The third level is Advanced. In this level, soldiers who hold advanced, senior-level positions (e.g., majors, lieutenant colonels, and sergeants) would develop shared reasoning, broaden their knowledge of various culture contexts, perceive world views in complex contexts, and be able to project cultural perspectives to others.

To facilitate the professional development of these soldiers, the author further introduces some realistic best practices, listed in Table 7.1.

FUTURE APPROACHES IN CULTURAL TRAINING

As Col. Maxi McFarland (retired, US Army) argued, the US military will likely continue to partner with other nations and regional powers as allies (2005). Further, with a significant percentage of vital natural resources (i.e., crude oil) under the control of foreign leaders, these regions remain important for national security. (Some estimates state that over 30% of oil imports come from regimes that are less friendly or stable, including Saudi Arabia, Venezuela, Nigeria, Angola, Iraq, and Algeria.) These two factors alone reinforce the argument that culture will continue to be important to the US military. However, research and on-the-ground experience have shown that it takes significant time and resources to increase the culture competence of American soldiers and related personnel.

One strategy for enhancing culture training is to start exposing promising soldiers to culture earlier in their careers. Currently, the United States Military Academy at West Point provides cadets opportunities for studying aboard. By interacting with and experiencing culture within a context and at a young age, the goal is to situate future officers in meaningful cultural settings that will speed learning and reinforce important cultural concepts. However, as McFarland suggested, such programs could be expanded into the Reserve Officer Training Corps, which could then partner with local universities in developing foreign exchange programs, such as the Fulbright Scholars Program. Additionally, the author discussed two areas of concern that must not be overlooked when implementing programs and training devoted to cultural awareness—namely, religion and tribal affiliation, which McFarland argued are worthy of future study for culture understanding.

It has been made painfully evident by the "War on Terror" that the importance of understanding an opponent's religion cannot be overstated. Former Secretary of State Madeleine Albright agreed that by attempting to understand the religious beliefs and traditions of other players on the global stage, international relations can be improved (Lawton, 2006). In fact, Albright further suggested that religious leaders should serve as advisors to high-ranking officials who interact with those of other religious traditions. In other words, to foster peace, military personnel must be equipped to understand the breadth and depth of various religious traditions within the context of the current geo-political outlook. In short, training programs must include meaningful opportunities to question, explore, and analyze diverse social aspects of religion.

According to McFarland (2005), culture is typically not well defined or bounded by written rules. Thus, to be effective, soldiers might need to rely on certain unwritten rules of conduct and behavior observed by local tribes. In fact, these unwritten tribal rules are often the only structures that are observed in "ungovernable" areas. By understanding these various collections of tribal rules (which would necessitate local language facility or complete reliance on a native speaker), the culture of a country could be learned, since it is often assembled from the shared understanding of diverse tribal rules.

To augment cultural understanding, the Defense Regional and Cultural Capabilities Assessment Working Group (WG) developed sets of guidelines for further integrating cultural factors into future planning and execution of operations (McDonald, McGuire, Johnston, Selmeski, and Abbe, 2008). Specifically, these guidelines can be categorized into three broad recommendations. First, standardize cultural training across all branches of services and agencies through a general culture-centric curriculum. Without such standardization, assessment will be a challenge. In fact, the lack of standardization is the main cause for the absence of assessment criteria and measurable outcomes that can then be addressed with programmatic changes. The WG also recommended that the US military look into how policies such as equal employment and diversity education are operationalized, since these practices also train for cross-cultural awareness and improved interpersonal interactions. Second, through an extensive research program, identify personal (e.g., motivation and emotion) and organizational factors that would enable military and civilian personnel to attain cultural knowledge in a cross-cultural environment. By identifying and understanding these factors through the input of experts in education, the private sector, and NGOs, effective training and professional development involving cultural awareness can then be established. Third, assessment metrics and tools must be developed and routinely evaluated to determine their ongoing validity for use in assessing culture training programs.

CONCLUSION

This review has substantiated the fact that cultural awareness will continue to be an important area for US national security forces. As the DoD's Quadrennial Defense Review reiterated, culture remains an important issue; as such, more resources should be made available for enhancing the language skills and culture training of our nation's military personnel (Quadrennial Defense Review Report, 2010). Additionally, the DoD proposes increasing the culture competencies of adjunct civilian personnel through the Human Terrain System (described in Section 1.3). In short, military decision makers must decide how to train relevant personnel in culture awareness so that operational outcomes may be improved.

The lack of standardization remains an issue of concern. As discussed in this review, there are numerous definitions and concepts related to culture (e.g., cultural awareness, cultural competence, cultural knowledge, cultural intelligence, cultural savvy, cultural capabilities, cultural expertise, cultural agility, cultural adaptability, cross-cultural skills, etc.) that are addressed in different ways depending on the DoD branch and who is doing the defining. For example, sometimes these definitions and concepts emerge from experienced soldiers in the field—and these may or may not correspond to those derived from internal or external "experts" charged with educating soldiers. Hence, decision makers may have difficulty prioritizing the myriad culture-related information that should be conveyed to military personnel before they are deployed. This lack of standardization is also reflected in the various schools and centers established by the different service branches—and this presents both advantages and disadvantages. On the one hand, these distinct schools and their curricula provide soldiers with unique perspectives, depending on what that particular branch feels should be taught to their personnel. On the other hand, different perspectives that lack a common set of underlying beliefs could introduce conflicting ideas about culture—especially since "joint operations" are becoming more important.

This lack of standardization about how to operationalize culture has resulted in two recommendations. First, the DoD should bring the various definitions and concepts of culture together into a single operationalized idea of culture to be promulgated across the various service branches. Second, a common cultural training program should be established across all service branches, while at the same time maintaining the different schools and centers that also teach more specialized culture concepts depending on the mission or need. Such a common cultural training program could facilitate the delivery of a basic cultural curriculum with learning objectives for all branches and ranks.

Another problematic issue with respect to culture training in the military is the lack of quantitative and rigorous studies designed to measure the effectiveness of cultural training and cultural awareness. This review has shown that most of the current studies and systems focus on subjective observations, policy goals, or thesis-related frameworks. In comparison, there was only one study based on quantitative objective analysis of survey data which was used to compare and identify soldiers' perspectives on culture. However, to determine the actual effectiveness of different culture training programs and systems offered by companies and service branch schools and centers, additional quantitative studies are needed so that objective conclusions can be made. Hence, one recommendation is to encourage the development of added studies to compare and contrast the various training programs. It should be noted, however, that the concept of "effectiveness" is an abstract and usually subjective concept. Therefore, applying Bloom's Taxonomy would be one approach to accomplish this recommendation for quantitative analysis. Bloom's Taxonomy is a set of foundational learning objectives that are widely used in the training and education domains (Bloom, 1956; Krathwohl, Bloom, and Masia, 1964). By operationalizing effectiveness according to the learning objectives of Bloom's Taxonomy, cultural training programs could then be compared and contrasted more objectively.

Lastly, a recent discussion (Fosher, personal communication, January 2011) with Kerry Fosher (the first command social scientist at the Marine Corps Intelligence Activity in Quantico, VA) and a review of two related publications (Fosher, 2008; Nuti, 2007) have led to two lessons learned. One lesson is that there is a need for anthropologists and DoD experts to continue collaborating and dialoguing on operationalizing relevant concepts of culture. This recommendation suggests that the skills and knowledge of anthropologists should be further leveraged for designing culture training programs and operationalizing concepts of culture. However, as Dr. Fosher reflected on her recent participation in the AAA Ad Hoc Commission on the Engagement of Anthropology with the US Security and Intelligence Communities, an anthropologist's code of ethics obliges that individual to serve his or her organization—while at the same time protecting the group or culture that is being studied and subsequently described to decision makers. This balancing act in and of itself points to the complexity of this issue—namely, that any culture and/or language training programs or devices should consider all aspect of a system so that local belief systems and cultural norms can be respected, while at the same time achieving whatever military goals are inherent in the mission.

In summary, this evaluation of the current literature on the importance of culture for contemporary military use reinforces that it is a complex issue—in part, because there is no definable benchmark for "success." Nonetheless, experts agree that attempting to effectively engage an opponent without knowing the cultural context of that individual or the region that person is defending would be like fighting with one arm tied behind one's back.

REFERENCES

Abrahams, D. (2007). Emotional Intelligence and Army Leadership: Give It to Me Straight! *Military Review,* March–April, 86–93.

Bloom, B. (1956). *Taxonomy of Educational Objectives, Handbook I: Cognitive Domain.* New York: Addison Wesley Publishing Company.

Crane, K., Oliker, O., Bensahel, N., Eaton, D., Gayton, J., Lawson, B., Martini, J., Nasir, J., Reyna, S., Parker, M., Sollinger, J., and Williams, K. (2009). Guidebook for Supporting Economic Development in Stability Operations. Retrieved 12/05/2010, from http://www.rand.org/pubs/technical_reports/2009/RAND_TR633.pdf

Erwin, S. (2009). Wanted: Soldiers With Cultural Savvy. Retrieved 01/17/2011, from http://www.nationaldefensemagazine.org/archive/2009/July/Pages/WantedSoldiersWith CulturalSavvy.aspx

Filkins, D., and Burns, J. (2006, May 1). The Rearch of War: Military; Deep in a U.S. Desert, Practicing to Face the Iraq Insurgency. *The New York Times.* Retrieved 07/22/2013 from http://query.nytimes.com/gst/fullpage.html?res=9DODE6DD113FF932A35756C0A96 09C8B63.

Fosher, K. (2008). Illuminating Competing Discourses in National Security Organizations. *Anthropology News, 49*(8), 54–55.

Gratch, J., and Marsella, S. (2001). *Tears and Fears: Modeling Emotions and Emotional Behaviors in Synthetic Agents.* Paper presented at the Proceedings of the 5th International Conference on Autonomous Agents, Montreal, Quebec, Canada.

Johnson, W. L., Beal, C., Fowles-Winkler, A., Lauper, U., Marsella, S., Narayanan, S. S., Papachristou, D., and Vilhjalmsson, H. (2004). *Tactical Language Training System: An Interim Report.* Paper presented at the Proceedings of the Conference on Intelligent Tutoring Systems (ITS), Berlin, Germany.

Kleiner, C. (2008). *Importance of Cultural Knowledge for Today's Warrior-Diplomats.* U.S. Army War College, Carlisle Barracks.

Krathwohl, D., Bloom, B., and Masia, B. (1964). *Taxonomy of Educational Objectives, Handbook II: Affective Domain (The Classification of Educational Goals).* New York: David McKay Company, Inc.

Lawton, K. (2006). Madeleine Albright. Retrieved 01/31/2011, from http://www.pbs.org/wnet/religionandethics/episodes/may-19-2006/madeleine-albright/1845/

Lewis, M. (2010). The Army's Foundational Building Blocks of Soldier Skills. Retrieved 01/01/2011, from http://www.army.mil/-news/2010/09/30/45902-the-armys-founda-tional-building-blocks-of-soldier-skills/index.html.

McCollum, C., Barba, C., Santarelli, T., and Deaton, J. (2004). *Applying a Cognitive Architecture to Control of Virtual Non-Player Characters.* Paper presented at the Proceedings of the 2004 Winter Simulation Conference, Washington, DC.

McDonald, D., McGuire, G., Johnston, J., Selmeski, B., and Abbe, A. (2008). Developing and Managing Cross-Cultural Competence within the Department of Defense: Recommendations for Learning and Assessment. Retrieved 01/29/2011, from http://www.deomi.org/CulturalReadiness/documents/RACCA_WG_SG2_Workshop_Report.pdf

McFarland, M. (2005). Military-Cultural Education. *Military Review,* March–April, 62–69.

McFate, M. (2005). The Military Utility of Understanding Adversary Culture. *Joint Force Quarterly, 38* (43), 42–48.

Meyer, M. (2009). A History of Socio-Cultural Intelligence and Research under the Occupation of Japan. Retrieved 01/03/2011, from http://www.strategicstudiesinstitute.army.mil/pdf-files/PUB914.pdf

National Security Act of 1947 (2007). Retrieved 07/22/2013 from http://www.intelligence.senate/gov/nsaact1947.pdf.

Norris, M. (2007). Women in Combat. Retrieved 01/26/2011, from http://www.npr.org/series/14964676/women-in-combat

Nuti, P. (2007). Reflecting Back on a Year of Debate with the Ad Hoc Commission. *Anthropology News, 48*(7), 3–4.

Petraeus, D. (2006). Learning Counterinsurgency: Observations from Soldiering in Iraq. *Military Review,* January–February, 2–12.

Quadrennial Defense Review Report. (2010). Retrieved 12/05/2010. from http://www.defense.gov/qdr/images/QDR_as_of_12Feb10_1000.pdf.

Selmeski, B. (2007). Military Cross-Cultural Competence: Core Concepts and Individual Development. Retrieved 01/23/2011, from http://www.culture.af.edu/PDF/SelmeskiMay2007.pdf

Stringer, M. (2008). *Women Training Iraqi Soldiers and Police in Iraq: The War with Three Fronts.* Air Command and Staff College.

Szalma, J., Hancock, P., and McDonald, D. (2009). Attitude of Military Personnel to the Challenge of Culture: Initial Findings. Paper presented at the 7th Biennial EO Diversity and Culture Research Symposium Patrick Air Force Base, Florida.

U.S. Navy Language Skills, Regional Expertise and Cultural Awareness Strategy. (2008). Retrieved from http://www.navy.mil/maritime/Signed_Navy_LREC%20Strategy.pdf.

Veroff, B., and Goldberger, R. (1995). What's in a name? The case for "intercultural." In *The Culture and Psychology Reader.* New York: New York University Press.

Watson, J. (2010). Language and Culture Training: Separate Paths? *Military Review,* March–April, 93–97.

Wolk, H. (1996). Toward Independence: The Emergence of the U.S. Air Force 1945-1947 [Electronic Version] from http://www.dtic.mil/cgi-bin/GetTRDoc?AD=ADA433273&Location=U2&doc=GetTRDoc.pdf.

Wunderle, W. (2006). *Through the Lens of Cultural Awareness: A Primer for US Armed Forces Deploying to Arab and Middle Eastern Countries.* Fort Leavenworth: Combat Studies Institute Press.

Zahn, M., and Lacey, W. (2007). *Building a Virtual Cultural Intelligence Community.* Monterey, CA: Naval Postgraduate School.

8 Disaster Management Systems

Rashaad E. T. Jones, Haydee M. Cuevas,
Cheryl A. Bolstad, Taylor J. Anderson,
Diana Horn, and Mica R. Endsley

CONTENTS

INTRODUCTION

Disaster management systems are designed to enhance the coordination and collaboration among public safety organizations that provide support for victims of natural disasters and other catastrophic events such as disease outbreaks. However, while designed with sophisticated capabilities to support disaster response efforts, the features and functions of these advanced technologies often do not address important cultural ergonomic issues that may hinder the performance of public safety personnel. In this chapter, we focus on two classes of cultural ergonomics issues that are relevant in disaster management: (1) meeting the collaboration and coordination needs of distributed teams with diverse backgrounds, such as nationality, culture, and ethnicity and (2) meeting the specific human performance needs of teams comprised of special populations, namely, older adults. We propose that these challenges can be addressed by applying sound human factors and human-computer interaction (HCI) principles, with a particular emphasis on promoting the *situation awareness* (SA) of individuals and teams. Our proposed solution involves adopting an SA-oriented design approach (SAOD) for developing disaster management systems with consideration for relevant cultural ergonomics issues.

We begin this chapter with a brief overview of situation awareness in individuals and teams, highlighting the critical role of SA in disaster management. We then describe our SAOD approach to creating user-centered systems, distinguishing between identifying users' information requirements and meeting their information demands. Two case studies are then presented to illustrate the application of our SAOD approach: (1) meeting the collaboration and coordination needs of Centers for Disease Control and Prevention (CDC) personnel and health professionals with diverse backgrounds and (2) meeting the specific human performance needs of American Red Cross volunteers, many of whom are older adults (age 65 and up). We conclude with implications for applying our SAOD approach to generate technology solutions to address other cultural ergonomics issues in complex domains.

SITUATION AWARENESS IN INDIVIDUALS AND TEAMS

Situation awareness involves being aware of what is happening around you to understand how information, events, and your own actions will affect your goals and objectives, both now and in the near future. More formally, SA can be defined as "the perception of elements in the environment within a volume of time and space, the comprehension of their meaning, and the projection of their status in the near future" (Endsley 1995, 36). As implied by this definition, SA is comprised of three levels: perception, comprehension, and projection (see Figure 8.1). Perception (Level 1 SA) involves an active process, whereby individuals extract significant cues from their environment, selectively directing attention to important information, while disregarding nonrelevant items. For example, to support Level 1 SA, disaster management systems can be designed to provide public safety personnel with information regarding the type of emergency and the number of people affected. Comprehension (Level 2 SA) involves integrating this information in working memory to understand how the information will influence the individual's goals and objectives. To support Level 2 SA in a severe emergency that affects several locations, disaster management systems can be designed to provide information that enables public safety personnel to understand the impact that assigning resources to a specific location will have on their ability to provide effective recovery and relief to people in all affected areas. Projection (Level 3 SA) involves extrapolating this information forward in time to determine how it will affect future states of the operating environment. In this case,

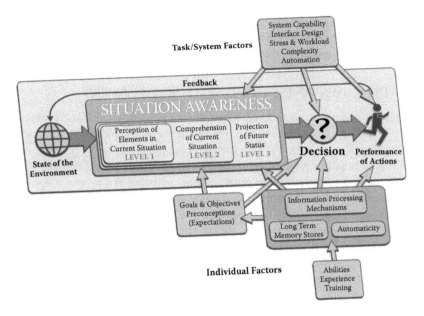

FIGURE 8.1   Theoretical model of situation awareness. (Adapted from Endsley, M. R. 1995. *Human Factors* 37(1): 32–64.)

disaster managements systems can be designed to provide trending information that enables public safety personnel to project the rate of growth of affected people for a given emergency.

TEAM AND SHARED SITUATION AWARENESS

Because in many complex domains individuals work as part of a team, the concepts of *team SA* and *shared SA* are as important as individual SA. Team SA can be defined as "the degree to which every team member possesses the SA required for his or her responsibilities" (Endsley 1995, p. 39). Thus, to ensure successful performance, each team member needs to have superior SA on those factors that are relevant for his or her job. In contrast, shared SA can be defined as "the degree to which team members possess the same SA on shared SA requirements" (Endsley and Jones 2001, p. 48). As implied by this definition, certain SA information requirements may be relevant to multiple team members. A major part of teamwork involves the area where these requirements overlap. In this case, successful team performance is influenced by the degree to which team members share a common understanding of what is happening on these shared SA elements. In other words, team members must be able to access and similarly interpret important information on the shared SA requirements that are relevant across their different positions.

To illustrate, when a disease outbreak occurs in multiple countries, physicians within each country need to know the disease symptoms, possible spreading mechanisms, and treatments that have been helpful. This necessitates shared SA across the different doctors regarding the disease. In contrast, government health officials have to know the number of cases within each country and possible links between these cases to determine possible trends. This requires team SA among different government personnel to gather the needed information.

SITUATION AWARENESS IN DISASTER MANAGEMENT

As depicted in Figure 8.1, situation awareness provides the foundation for subsequent decision making and performance in the operation of complex and dynamic work environments and systems (Endsley 1995). SA encompasses not only the perception of critical information within the environment, but an increased understanding of that information such that future events can be predicted and proper action can be taken. SA is especially important in disaster management, both at the individual and team level. To illustrate, the first step in a disaster management is to identify whether a real emergency is present, the type of emergency, and the number of people affected (Level 1 SA—perception). Here, what matters most is that public safety personnel are provided with and are able to attend to the correct data and information. However, performance at this stage can by influenced by individual differences in training and experience (e.g., novice–expert differences), as well as by the design of the systems available to provide information relating to the emergency; these factors can both directly and indirectly influence what information is perceived. For example, experts, as compared to novices, are likely to have developed a greater sensitivity for detecting or recognizing patterns in specific types of data through training, extensive experience in conducting relief efforts, better focused attention, or more effective use of data representations (Garrett and Caldwell 2009). In turn, this greater sensitivity may enable them to recognize relevant information with more accuracy and speed.

The second step in disaster management requires understanding patterns or trends in the data as well as evaluating the availability of resources to meet the challenge (Level 2 SA—comprehension). For instance, public safety personnel need to be able to determine whether a temporary shortage in resources (e.g., personnel, food, water) will significantly hinder their ability to respond effectively to the current demands of the emergency. Finally, the third step involves predicting future trends and the ability to meet the continuing demands of the emergency (Level 3 SA—projection). This includes conducting "what-if" analyses involving social, organizational, economical, environmental, and political trends. For example, if an emergency affects multiple geographic areas (e.g., widespread flooding and windstorm damage caused by a hurricane), public safety personnel will need to notify other state and national agencies to help respond to the projected demands of the emergency.

Endsley's theoretical model of SA also illustrates several variables that can influence the development and maintenance of SA in disaster management. SA may be affected by the inherent complexity of the tasks involved in monitoring and responding to emergencies as well as the design of systems available for monitoring these public safety threats. Public safety personnel are often bombarded with too much data, but not enough reliable and interpretable information that they can use to guide their decision making (as noted in the 2010 Haiti earthquake). Adding to this complexity is that sources of information are not always obvious and disaster management systems do not capture "soft" data such as expert opinions. In addition, natural disasters and other public safety threats (e.g., disease outbreaks) can evolve and change over time, as evident in the New Orleans flood following Hurricane Katrina in 2005 and the 2009 H1N1 swine flu outbreak. As such, disaster management systems must be designed to present information so the human user can quickly recognize and understand the impact of the emergency and forecast how the emergency will change over time.

Disaster management involves many people from different disciplines and agencies, with unique roles and a varying range of experience, working together to monitor and respond to potential public safety threats, both locally and nationally and, in some cases, internationally. In addition, teams are often distributed (e.g., some personnel are in the field while others are in a headquarters facility), which presents challenges for effective communication and coordination. Thus, disaster management systems need to go beyond keeping one human in the loop (individual SA) but, more importantly, support collaboration and sharing of information across organizations and teams composed of culturally and demographically diverse members (team and shared SA). However, as noted earlier, individual factors may affect the SA of public safety personnel, including their diverse background, training, and experience. As will be discussed next, our SAOD approach addresses these challenges to individual and team SA by applying sound human factors and HCI principles to the design of disaster management systems with consideration for relevant cultural ergonomics issues.

SITUATION AWARENESS-ORIENTED DESIGN (SAOD)

Our approach is based on the SAOD process developed by Endsley and colleagues (see Endsley, Bolte, and Jones 2003) as a means to improve human decision making and performance in complex dynamic environments through optimizing individual and team SA. The SAOD process is user-centered and derived from a detailed analysis of the goals, decisions, and SA requirements of the operator. This method has been successfully applied as a design philosophy for complex systems involving remote maintenance operations, medical systems, flexible manufacturing cells, and military command and control. The SAOD process consists of three main components: SA requirements analysis, SAOD principles, and SA measurement and validation. SA requirements analysis utilizes Goal-Directed Task Analysis (GDTA), a unique cognitive task analysis methodology that focuses not only on what data the operator needs, but also on how that information is integrated or combined to address each decision. SAOD Principles include a set of 50 design principles based on a theoretical model of the mechanisms and processes involved in acquiring and maintaining SA in dynamic complex systems. Finally, SA measurement and validation is critical during the design process if optimizing SA is to be a design objective. Without this evaluative component, it will be impossible to tell if a proposed concept actually facilitates SA, has no effect, or inadvertently hinders SA in some way. Two components of the SAOD process, SA requirements analysis and SAOD principles, are especially relevant to disaster management systems. These are described in greater detail next.

SA REQUIREMENTS ANALYSIS

The SAOD process begins by first identifying the critical SA requirements for the targeted position. SA requirements are defined as those dynamic information needs associated with the major goals or subgoals of the operator in performing his or her job (as opposed to more static knowledge such as rules, procedures, and general system knowledge). These critical SA requirements can by identified utilizing the GDTA methodology, which involves conducting extensive knowledge elicitation sessions with domain subject matter experts, observation of performance of tasks, and analysis of written materials and documentation (for a detailed description of this methodology, see Endsley et al. 2003). The objective of the GDTA is to identify the major goals and decisions that drive performance in a particular job or position as well as to delineate the critical, dynamic information requirements associated with each goal and decision. By focusing on goals rather than tasks, this methodology seeks to identify information needs directly without considering *how* the user will acquire the information. Thus, the GDTA not only provides a technology-independent means to identify the specific information the user requires to accomplish a specific goal, but also indicates the manner in which the user integrates this information in order to develop a robust understanding of the current situation and to project how the situation will change over time.

The output of the GDTA graphically illustrates the hierarchical relationship between the various goals, subgoals, decisions, and SA requirements (see Figure 8.2). Although a numbering convention is used to show associations between goals and subgoals; these numbers are used for documentation purposes only and do not imply any sequencing or goal prioritization, as priorities vary over time and some subgoals or decisions may be active or inactive at any point in time. Primary goals and subgoals are presented above the decision(s) associated with each subgoal. Information requirements for the three levels of SA (perception, comprehension, and projection) relevant to each decision are organized to illustrate the manner in which lower-level information (Level 1 SA) is integrated to address higher-level SA requirements (Levels 2 and 3 SA).

SAOD PRINCIPLES

SA Requirements Analysis provides insight into operators' critical *information requirements*, that is, *what* information needs to be displayed. In turn, SAOD Principles provide guidance on meeting the operators' *information demands*, that is, *how* this information should be presented to support their SA. SAOD Principles are focused on a model of human cognition involving dynamic switching between goal-driven and data-driven processing and feature support for limited operator resources. These 50 principles are organized into six categories (see Figure 8.3). Applying these SAOD Principles ensures that systems (1) present the key information to promote SA development and maintenance, (2) integrate information to support comprehension and projection of the status of ongoing operations, (3) provide "big picture" information to promote global SA, while enabling easy access to details needed for situation understanding, (4) use information salience to direct the operator's attention to

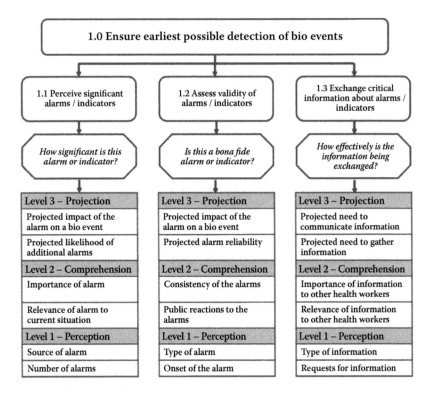

FIGURE 8.2 Illustrative example of GDTA output for public health professionals responsible for detection and management of potential disease outbreaks. (Adapted from Bolstad, C. A. et al. 2011. *Biosurveillance: Methods and Case Studies*, ed. T. Kass-Hout and X. Zhang, 79–94. Boca Raton, FL: CRC Press–Taylor & Francis Group.)

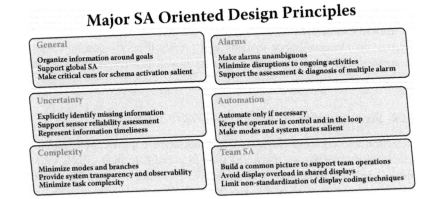

FIGURE 8.3 Examples of SA-Oriented Design Principles.

important information and events, and (5) directly support multitasking activities that are critical for SA.

Our principled SAOD approach can guide the design of culturally appropriate disaster management systems to ensure that public safety personnel are able to provide effective services (cf. Kaplan 1995). Next, we will demonstrate the application of our approach with two case studies. In each case study, we briefly describe the domain, highlighting important cultural ergonomics issues. We then explain how we applied two components of the SAOD process to generate SA-oriented technology solutions to address these issues.

CASE STUDY #1: CDC BIOSURVEILLANCE TEAMS

OVERVIEW OF DOMAIN

The CDC is dedicated to creating the expertise, information, and tools that people and communities need to protect public health, such as through prevention of disease and preparedness for new health threats (CDC 2010). CDC personnel are integrally involved in *biosurveillance*, which refers to the process of detecting and categorizing diseases and disease outbreaks in people, animals, or plants by monitoring elements in the environment that may cause these diseases. To support its biosurveillance operations, the CDC's Office of Critical Information Integration and Exchange (OCIIX) initiated the BioPHusion program to facilitate the integration or *fusion* of public health information from existing CDC programs, external partners, and other sources to increase situation awareness, improve early event detection, and support decision making to protect the public's health (Rolka, Walker, and Heitgerd 2009). Its scope encompasses sharing timely and actionable information to public health programs and leaders at the local/tribal, state, national, and global levels. CDC BioPHusion personnel rely on *biosurveillance systems*, a unique type of disaster management system, designed to collect existing health-related data and analyze this information for the purpose of detecting disease cases, disease outbreaks, and environmental conditions that increase susceptibility to diseases (e.g., contaminated drinking water; Fricker, Hegler, and Dunfee 2008). Biosurveillance systems that can support early detection and real-time interpretation of information represent a critical front line of defense for modern epidemic and disease control. However, the effectiveness of biosurveillance systems will greatly depend upon how well their design supports their user's situation awareness as well as addresses relevant cultural ergonomics issues.

CULTURAL ERGONOMICS ISSUES IN BIOSURVEILLANCE TEAMS

The detection and control of disease outbreaks often requires the creation of *ad hoc* teams interacting across space and time to coordinate their efforts to achieve a common goal. These teams are composed of individuals with a broad range of specializations within distinct domains (e.g., epidemiologists, physicians, nurses, veterinarians, computer scientists, statisticians, water quality specialists, biologists, and microbiologists) and are drawn from different local, state, and federal health agencies as well as international organizations (Wagner, Moore, and Aryel 2006). The dynamic, fluid, multidisciplinary, and often multicultural nature of these ad hoc teams may influence their shared SA, and in turn, their ability to collaborate and coordinate to achieve their goals. For example, ad hoc teams may have unique shared SA requirements that differ from those of intact groups accustomed to working together, such as specific information regarding their team members' knowledge, skills, and abilities needed to guide assignment of roles and responsibilities (Strater et al. 2009).

Consequently, relevant cultural ergonomics issues in this domain include how to design biosurveillance systems that support the situation awareness and decision making performance of biosurveillance personnel and other public health professionals with varied cultural background, training, and experience. An effective biosurveillance system needs to synthesize the information available to provide an accurate common picture of the situation for all personnel who will access and use such a system, thereby ensuring shared SA across team members. Another major challenge in biosurveillance is that the sources of data are infinite. While this creates the need for biosurveillance systems to include a broad range of information sources, the key to maximizing operational efficiency is to determine what information is significant and most useful. Next, we describe our research efforts aimed at addressing these challenges facing biosurveillance personnel in the CDC BioPHusion Center.

SA REQUIREMENTS ANALYSIS—IDENTIFYING INFORMATION REQUIREMENTS

As noted earlier, the BioPHusion program requires effective collaboration and coordination across numerous organizations and agencies involving many individuals. A necessary first step in our research in this domain, therefore, involved identifying the critical SA requirements of these public health professionals. We conducted GDTA interviews with 26 subject matter experts drawn from the following programs: Coordinating Center for Infectious Disease, Coordinating Office for Global Health, Coordinating Office for Terrorism Preparedness and Emergency Response, Coordinating Center for Environmental Health and Injury Prevention, Influenza Coordinating Unit, National Center for Preparedness, Detection, and Control of Infectious Diseases National Center for Immunizations and Respiratory Diseases, and Northrop Grumman Corporation (Medical Division). Findings from these interviews were compiled and integrated to create an overarching GDTA that represented the BioPHusion Center as a whole. The complete GDTA is too detailed and comprehensive to present in this chapter (over 20 pages in length). Thus, our discussion will focus on relevant portions from the BioPHusion Center GDTA that highlight important team coordination and cultural ergonomics issues.

To begin, the main goal for BioPHusion biosurveillance personnel is to "Take actions needed for significant public health events." This goal is achieved by accomplishing five subgoals:

1.0 Identify significant public health events.
2.0 Exchange critical public health information.
3.0 Determine actions needed for response to significant public health events.

4.0   Track effectiveness of response to current significant public health events.

5.0   Assess preparedness for significant public health events.

The decision and some of the corresponding information requirements associated with subgoal 2.0 are presented in Table 8.1. Note that these examples are only a very small sampling of the many SA requirements associated with this subgoal. Nevertheless, this GDTA excerpt provides insights into the operational challenges faced by biosurveillance teams. For example, to support higher levels of SA (comprehension and projection), biosurveillance systems must not only provide details about a potential disease outbreak, but also support users in determining the reliability of and confidence in the information and its timeliness. Unreliable, uncertain, and/or outdated information could lead to poor decision making regarding the severity of a public health threat, resulting in delayed detection and response and potent loss of human life. In addition, these information requirements also draw attention to cultural ergonomics issues that may influence biosurveillance team performance. Specifically, the perceived trustworthiness of the source of the communication can be both directly and indirectly affected by salient cultural or ethnic differences

---

TABLE 8.1

Excerpt from BioPHusion Center GDTA

Subgoal: 2.0 Exchange critical public health information

Decision: How effectively is public health information being exchanged?

- L3 Projected level of confidence in communications
  - L2 Impact of trust in communication on confidence in communications
    - L1 Source
  - L2 Impact of communication understanding on confidence in communications
    - L1 Complexity
    - L1 Clarity
  - L2 Impact of information on confidence in communications
    - L1 Reliability
    - L1 Source (internal, external)
    - L1 Confidence
    - L1 Consistency
    - L1 Format
  - L2 Impact of criticality of information on confidence in communications
    - L1 Information priority
    - L1 Relevance
  - L2 Impact of information timeliness on confidence in communications
    - L1 Time of event
    - L1 Time event information is received
    - L1 Time event information is sent

Note: L3 = Level 3 SA (projection); L2 = Level 2 SA (comprehension); L1 = Level 1 SA (perception).

---

between the sender and recipient of the information. Direct influences may stem from biases regarding the perceived competence of individuals from different cultural or ethnic backgrounds or the trustworthiness of the systems they are using to report data. Indirect influences include potential misunderstandings and breakdowns in communication due to team members speaking different languages. Thus, to be truly effective, the design of biosurveillance systems must also consider the cultural ergonomics issues that may potentially interfere with biosurveillance personnel's SA and task performance.

SA-ORIENTED DESIGN PRINCIPLES—ADDRESSING INFORMATION DEMANDS

Biosurveillance systems that enhance operators' awareness of what is happening in a particular situation can dramatically improve their ability to detect and manage potential disease outbreak. A well-designed system should simultaneously focus on providing critical data that is structured to support the active goals of the end user while at the same time promoting team and shared SA. As noted in the preceding section, the GDTA methodology can be used isolate this essential data by identifying operators' critical information requirements. In turn, applying relevant SAOD Principles will help ensure that biosurveillance systems are designed to address operators' information demands, that is, optimally present this information to its users. Because of the limited length of this chapter, we will focus on briefly reviewing only a few key design principles for supporting SA in biosurveillance operations. (For a more detailed discussion, see Bolstad, Cuevas, Wang-Costello, Endsley, Page, and Kass-Hout 2011.)

Organize information around goals. This principle emphasizes that effective systems must provide information organized around the operator's major goals and subgoals, as identified in the GDTA. Although it may seem intuitive to create a system with separate modules based on the operator's main goals, this would not lead to a very usable interface as some functions would be buried under several menu layers. A better approach would be to design a user interface that utilizes tabs as shown to present information requirements related to the operator's subgoals at a top level:

Present Level 2 SA information directly to support comprehension. System design needs to support interpreting and comprehending all incoming biosurveillance data. This includes mapping and interpretation of alerts, trend data, and environmental and organizational factors as well as being able to distinguish between false alarms and undiagnosed or unconfirmed events.

Provide assistance for Level 3 SA projections. Biosurveillance systems need to not only present data to support lower levels of SA (detection of a possible outbreak and understanding patterns or trends in the data), but, more importantly, higher levels of SA to determine the future impact of the information (predicting the future trends and distribution of the outbreak).

Represent information timeliness. One of the problems facing biosurveillance personnel is information recency, that is, the timeliness of the information presented. Thus, biosurveillance systems must provide support for enabling users to recalculate the current picture when they receive late data.

Biosurveillance systems must also enable users to distinguish between new and past information and be alerted to the data's age, as certain diseases can spread so rapidly that even a few hours can significantly reduce the data's usefulness for predicting trends.

Build a common operating picture to support team operations. This principle is especially relevant for addressing the cultural ergonomics issues associated with multidisciplinary, multicultural teams, as the diverse background, training, and experience of team members can hinder the development of shared SA. To mitigate these issues, biosurveillance systems can provide shared information displays and virtual collaborative spaces to help distributed team members establish a shared understanding of the situation. Even providing a standard mapping tool such as Google Maps will help to build shared SA.

Support transmission of different comprehensions and projections across teams. Tools need to allow sharing of intuitive assessments between biosurveillance personnel and other public health professionals. Currently, data analysis and detection tools are not integrated with listserves and other collaboration tools to support assessment sharing. Yet, understanding individual interpretations and team consensus is paramount to effective SA and decision-making performance.

TECHNOLOGY SOLUTION—SA-ORIENTED BIOSURVEILLANCE SYSTEM

Figure 8.4 illustrates how some of these SAOD Principles were applied to create the user interface of our prototype SA-oriented biosurveillance system to support both individual and team SA. Tabs representing four general work areas (Overview, Timeline, Event, Media) are used to enable quick and easy navigation to critical SA requirements associated with specific subgoals. The top-level screen shows, both graphically and in tabular format, all current events filtered by user-defined parameters (e.g., region, event type, event criticality). The Event Overview Table provides detailed information on each event, which is needed to determine its criticality and recency (Level 2 SA). Users can click on the View icons to access trending information needed to support Level 3 SA. Finally, the simplicity of the interface design helps promote team and shared SA, as team members can readily access the same information with ease and communicate this information in a similar fashion.

Our prototype SA-oriented biosurveillance system provides a common operating picture designed to help public health professionals gain a clear understanding of the current status of concurrent health-related events. This is particularly important when information providers and decision-makers are distributed across organizations, states, or even countries, and thus, may have different mental models of the situation. In particular, health officials from different cultures, or even organizations, tend to view data according to their own mental model and thus may lose sight of the common global health outlook. Our SAOD approach to system interface design supports this cultural diversity by integrating information into a common format, making Level 2 and 3 SA information more explicit, and providing

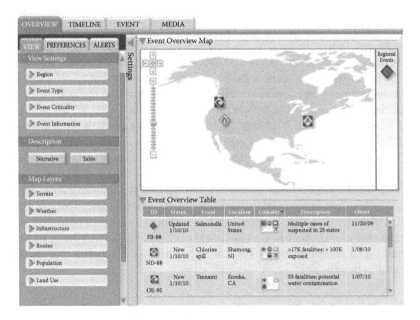

FIGURE 8.4 Illustrative example of our prototype SA-oriented biosurveillance system user interface.

high-level summaries that emphasize what is important and where critical resources need to be directed.

As noted earlier, biosurveillance systems must integrate data from many different sources, each with their own inherent reliability levels and limitations. To address this issue, our prototype SA-oriented biosurveillance system provides high-level information for each significant public health event and allows the user to drill down into specific event details, such as information source; this design feature supports the decision-maker in determining how best to interpret and utilize the available information. For instance, if a source is known to be unreliable, the decision-maker can confirm the reported information through other sources. Additionally, our SA-oriented biosurveillance system provides more salient Level 2 and 3 SA information, thereby supporting not just perception of elements, but also comprehension and projection of the situation. For example, the user interface of our system provides information details with regard to each event's impact on health, medical personnel, and resources, media, politics, national security, and trade and travel using consistent iconic images. Such goal-directed, detailed information enables decision-makers to effectively direct resources where most needed.

CASE STUDY #2: AMERICAN RED CROSS DISASTER ACTION TEAMS

OVERVIEW OF DOMAIN

The American Red Cross is a primarily volunteer-based humanitarian organization that provides support and services to people who are affected by natural disasters (e.g., hurricanes, floods, earthquakes, fires) as well as manmade disasters (e.g., hazardous materials spills, transportation accidents, explosions). Its disaster relief operations are supported by Disaster Services Human Resources personnel, which are responsible for major disasters that affect large geographical areas, and Disaster Action Teams, which are responsible for local emergencies that affect a few families, such as fire or water damage to a home. Our investigation focused on the latter group. Team members that volunteer for Disaster Action Teams are on call from 8 pm to 8 am, seven days a week and are responsible for providing immediate assistance (from 0 to 48 hours) to meet the emergency needs of individuals and families who are victims (referred to as "clients" by the American Red Cross) of natural and man-made disasters. Team members are dispatched after receiving an initial call from the fire department, disaster victim, family member or friend, and will respond to disasters in teams of two or more people. Once on-site, they evaluate the extent of damage to determine how much assistance clients will need. Clients are given the means to obtain essential items such as food and clothing, basic household appliances and furnishings, and medical and occupational supplies that are lost or destroyed in the disaster.

CULTURAL ERGONOMICS ISSUES IN DISASTER ACTION TEAMS

Disaster Action Teams face challenges similar to those experienced by biosurveillance teams. These include how to effectively coordinate the efforts of team members with diverse background, training, experience, and levels of expertise as well as how to best support collaboration among distributed organizations and agencies. As such, many of the issues discussed in Case Study 1 are relevant here as well. However, in our research with the American Red Cross (see Jones, Mossey, and Endsley 2010), we also identified two additional challenges for Disaster Action Teams. The first involves effectively coordinating and dispatching resources. The challenge is to utilize team members strategically so that the proper response is provided to all clients who have been affected by an emergency or disaster. This includes arriving at a location in a timely manner and providing the appropriate type of assistance. Effective resource management also requires utilizing team members efficiently to maintain an appropriate workload balance as well as to increase the overall knowledge base and experience of the entire team by exposing team members to a variety of disasters and emergencies.

The second challenge facing Disaster Actions Teams stems from the fact that a high number of volunteer team members are older adults (age 65 and above), with the majority over 45 years old. This creates important cultural ergonomics issues related to the specific human performance needs of this population. This involves consideration of potential cognitive and physical limitations associated with the effects of aging and age-related illnesses as well as difficulties in interacting with technology due to generational gaps. The cognitive and physical signs of aging begin to be noticeable as people reach their mid-forties, making the design of technology aimed at the 45 and older age group more challenging (Hawthorn 2000). With regard to generation issues, many older adults began using modern technology later in life and are not as proficient as younger generations who began using computers at a relatively young age (Nielsen 2002). Consequently, because of their unfamiliarity with modern user interfaces, older adults may require more time to learn and utilize computer-based technology (Watanabe, Yonemura, and Asano 2009).

Only a few notable research studies have investigated situation awareness in older adults. For example, Bolstad (2001) examined how the ability to attend to important information during driving changes with age and how this is related to normative aging changes in other cognitive abilities (e.g., useful field of view, dynamic memory, perceptual speed, and time-sharing). Aging-related changes in cognitive abilities that affect attention and memory may result in older adults experiencing difficulties in developing and maintaining SA and performing certain essential tasks in work-related settings (Bolstad and Hess 2000). Thus, a better understanding of the specific human performance needs of this special population can lead to the design of systems that enable improved SA and task performance in older adults. We next describe our research efforts aimed at addressing the unique challenges facing Disaster Action Teams.

SA Requirements Analysis—Identifying Information Requirements

While different Red Cross chapters may have slightly different organizational structures, in general, a single Disaster Action Team will have one captain that is responsible for several teams. Each team includes an assistant captain, a team leader, and several standard team members. While individuals assigned to the upper roles will generally have several years of experience with the American Red Cross, the experience levels of standard team members vary from new trainees to individuals with 5+ years of experience. To identify the SA requirements for Disaster Action Teams, we conducted a series of extensive GDTA interviews with a subject matter expert who was highly experienced in disaster relief operations. This expert has volunteered with the American Red Cross since 2005, supporting disaster relief operations in New Orleans after Hurricane Katrina. He has served as team lead, assistant captain, and captain of Disaster Action Teams. In particular, in 2009 and 2010, he served as the captain of the Northwest Atlanta Disaster Action Team, responsible for 150 volunteers providing services over an area of six counties, encompassing a population of 1.2 million people.

Because he has extensive experience as a Disaster Action Team captain, assistant captain, and team lead, this American Red Cross expert was able to provide detailed information on the responsibilities and SA requirements for all three roles of interest. From these interviews, we developed three GDTA hierarchies, one for each role. The complete Disaster Action Team GDTAs are too detailed and comprehensive to present in this chapter. Thus, we will discuss relevant portions from each of the GDTAs to highlight important team coordination and cultural ergonomics issues. Table 8.2 presents examples of the subgoals, decisions, and information requirements for the captain, assistant captain, and team lead.

Captains are responsible for organizing, training, equipping and scheduling teams to ensure that their chapter can provide effective relief services to people affected by emergencies. Thus, the main goal for the captain is to "Ensure efficient distribution of team resources." This goal is achieved by accomplishing two subgoals: "1.0

TABLE 8.2
Excerpt from Disaster Action Team GDTAs

Captain

Subgoal: 1.0 Effectively balance and load teams

Decision: How can teams be arranged and balanced to cover entire geographic sector?

- L3 Projected impact of team arrangement on response time
  - L2 Impact of member's geographic location on team assignment
    - L1 Residence location of volunteer and other volunteers
    - L1 Ranking
    - L1 Trends in incident locations
    - L1 Red Cross districts
    - L1 Population distribution of districts
- L3 Projected team assignment
  - L2 Impact of team cohesiveness on modification of team structure
    - L1 History of team performance
    - L1 History of team cooperation
    - L1 Individual reputation to respond

Assistant Captain

Subgoal: 2.0 Provide effective communication to dispatched team

Decision: What does the team leader need to know?

- L3 Projected impact of information on preparedness of team
  - L2 Impact of incident information on team preparedness
    - L1 Location
    - L1 Disaster type
    - L1 Client name
    - L1 First responder
    - L1 Case number
  - L2 Impact of information from dispatch on team preparedness
    - L1 Location
    - L1 Disaster type
    - L1 Phone number of incident commander
  - L2 Impact of information from incident commander
    - L1 Disaster size
    - L1 Expected number of clients
    - L1 First responders' location
    - L1 Client location

Team Lead

Subgoal: 1.0 Ensure effective interactions with clients

Decision: How can the team be most effectively used to ensure the best client interactions?

- L3: Projected abilities of team members
  - L2: Impact of individual skills
    - L1 Language
    - L1 Specialties

TABLE 8.2 (*Continued*)
Excerpt from Disaster Action Team GDTAs

  - L1 Experience
  - L1 Recency of last event attended
- L2: Impact of incident type on abilities of team members
  - L1 Type of home
  - L1 Number of homes
  - L1 Number of clients
  - L1 Family dynamic
  - L1 Cultural dynamic

Note: L3 = Level 3 SA (projection); L2 = Level 2 SA (comprehension); L1 = Level 1 SA (perception).

Effectively balance and load teams" and "2.0 Effectively encourage team members to pursue leadership roles." Table 8.2 lists the decision and corresponding information requirements associated with subgoal 1.0. To support this subgoal, captains would benefit from useful disaster management systems that enable them to quickly access and interpret these information requirements so that they can strategically and logistically assign team members to a specific team. The GDTA results for the captain also highlight the potential effect of cultural ergonomics issues related to team composition (e.g., culture, ethnicity) on team cohesiveness.

Assistant captains are directly responsible for dispatching resources to an emergency site. Thus, the main goal for assistant captains is to "Provide effective dispatch and coordination." This goal is achieved by accomplishing two subgoals: "1.0 Determine optimal team arrangement" and "2.0 Provide effective communication to dispatched team." Table 8.2 lists the decision and corresponding information requirements associated with subgoal 2.0. The decision for this subgoal highlights the importance of designing disaster management systems that support not only individual but also shared SA. Specifically, the information requirements for this assistant caption subgoal also directly influence the SA of the team lead. In addition, the assistant captain and captain also have similar shared SA requirements in that both are tasked with assigning members to teams to optimize the use of resources and ensure effective disaster response.

The team lead is the Disaster Action Team member that has the most effect on addressing clients' needs. Thus, it is not surprising that the team lead's main goal is to "Provide optimal client services." This goal is achieved by accomplishing three sub-goals: "1.0 Ensure effective interactions with clients," "2.0 Determine navigation pathways," and "3.0 Provide proper level of support." Table 8.2 lists the decision and some of the corresponding information requirements associated with the subgoal 1.0. These information requirements again highlight how team member differences in background, training, and experience must be considered when designing disaster management systems. The information requirements associated with subgoal 1.0 also draw attention to another cultural ergonomics issue relevant to disaster management. Specifically, thus far in this chapter we have focused on how team diversity may influence coordination within teams. However, team leads also face potential cultural ergonomics issues related to the cultural diversity of the clients they serve. They must achieve an understanding of how family and cultural dynamics may affect their ability to interact effectively with their clients.

SA-Oriented Design Principles—Addressing Information Demands

Several SA-Oriented Design Principles offer useful guidance for designing user interfaces for Disaster Action Teams that address team members' information demands by effectively presenting critical information requirements. These can be summarized as follows:

- Direct presentation of higher-level SA requirements (comprehension and projection) is recommended, rather than supplying only low-level data that operators must integrate and interpret manually.
- Goal-oriented information displays should be provided, organized so that information requirements for a particular goal are colocated and directly answer the major decisions associated with that goal.
- Support for global SA is critical. Information displays should provide an overview of the situation across the operator's goals at all times (with detailed information for subgoals of current interest) and enable efficient and timely goal-switching and projection.
- Essential cues related to key features of schemata should be made salient in the interface design. In particular, those cues that will indicate the presence of prototypical situations will be of prime importance and will facilitate goal-switching in critical conditions.
- Extraneous information not related to SA needs should be removed from the display, while carefully ensuring that such information is not needed for broader SA needs.
- Support for parallel processing, such as multimodal displays, should be provided in data rich environments.

Applying these principles to the design of the disaster management systems used by Disaster Action Teams will help ensure that team members can develop and maintain SA, and thereby enhance their ability to coordinate their efforts to effectively respond to an emergency. However, because designing SA-oriented systems for Disaster Action Teams must also consider cultural ergonomics issues associated with older adults, our SAOD approach involved examining how aging-related factors influence task performance, as discussed next.

Human Factors and HCI Design Guidelines for Older Adults

As part of our research for the American Red Cross, we conducted a detailed analysis on how aging-related decrements in vision, motor response, and cognitive functioning affect user interactions with computer-based technology (for a detailed discussion, see Jones et al. 2010). For example, as people age, their vision declines for a variety of reasons, including stroke, glaucoma, presbyopia, cataracts, and the shrinking of the pupil (AgeLight 2001). This makes it more difficult for older adults to read text on displays that use smaller fonts and have poor contrast. Similarly, speed of physical movements and accuracy of hand–eye coordination generally decrease with age, thus making the use of peripheral devices (e.g., mouse or track-ball) to select a button on the screen or scroll through a web page more challenging for older adults (Hawthorn 2000; Hollinworth 2009). With regard to cognitive functioning, aging-related reductions in processing resources (e.g., memory and attention) may limit the amount and complexity of information that older adults can process at any given time (Bolstad and Hess 2000). For example, decreased short-term memory capacity can cause older adults to experience problems when navigating webpages, such as losing their sense of "Internet orientation" and direction and forgetting the specific web page they recently visited or the hyperlinks that they clicked (AgeLight 2001; Nielsen 2002). The findings from our analysis were compiled and integrated to generate a set of human factors and HCI design guidelines to address specific aging-related issues that may hinder the task performance of older adults (see Table 8.3).

TECHNOLOGY SOLUTION—PINPOINT™

Insights gleaned from our SA Requirements Analysis, coupled with relevant SA-Oriented Design Principles and human factors and HCI design guidelines, were put into practice to inform the design of PinPoint™. PinPoint is an SA-oriented web-based (cloud computing) tool that is designed to support Disaster Action Team members in their preparation and response to disasters by effectively enhancing their coordination, supporting information sharing, and facilitating communication. As its name implies, our tool "pinpoints" resources, locations, activities, and information. PinPoint includes Google Mapping Services to display geospatial information, such as the location of events, which supports the Captain's goal of managing members strategically (Figure 8.5). PinPoint also supports navigational tasks, such as providing directions to an ongoing emergency. PinPoint also addresses the unique operational challenges faced by Disaster Action Teams by providing a system that is robust, streamlined, and flexible. It allows for resource coordination among constantly changing personnel and roles by enabling immediate sharing of critical information, providing information only when needed, and facilitating communication between resources. In addition, PinPoint is easy to use and requires minimal training, thereby increasing its utility for older adults. Incorporating design principles, such as utilizing action words for buttons and minimizing scrolling, further ensures efficient use by older adults. Initial feedback from American Red Cross personnel has been overwhelmingly positive, specifically with regard to the simplicity and ease of use of PinPoint's design. Our future research and development plans include conducting usability studies to evaluate the efficacy of PinPoint for supporting Disaster Action Teams in a realistic operational setting.

TABLE 8.3

Key Human Factors and HCI Design Guidelines to Address Aging-Related Issues

Aging-Related Issue	Design Guideline
Visual	*Colors and Illumination*
	• Use complementary colors
	• Colors should be bright and bold
	• Colors, such as black, white, blue, and yellow are preferable
	• Avoid flashing graphics
	Text
	• Avoid shading text and placing text on patterned backgrounds
	• Text color (foreground color) should be dark with a light background
	• Use 12-point font for regular text
	• Headlines should be 2 points larger
	• Use San Serif font
	• Provide easy way to increase font size
	• Only use bold text in appropriate places
Motor	*Buttons*
	• Use buttons that are relatively larger than the normal size typically used
	• Buttons and hyperlinks should be large and readily selectable
	• Use sufficient space between buttons and hyperlinks
Cognitive	*Terminology*
	• Use instructional action phrases for buttons
	• Avoid technical terms
	• Use forgiving and corrective feedback messages
	• Use short phrases
	• *Search*
	• Use effective search engines that examines the entire website
Motor/Cognitive	*Navigation*
	• Eliminate scrolling by placing all information on the visible portion of the screen
	• Avoid using drop-down and nested menus
	• Use blue color and underline hyperlink text and set the color to purple if the user has already clicked it
	• Use tabs, when appropriate

Source: Adapted from Jones, R. et al. 2010. Website usability design principles for the elderly. In *Proceedings of the Third International Conference on Applied Human Factors and Ergonomics (AHFE)*, ed. G. Salvendy and W. Karwowski, 833–841, July 17–20, 2010, in Miami, FL. Boca Raton, FL: Taylor & Francis.

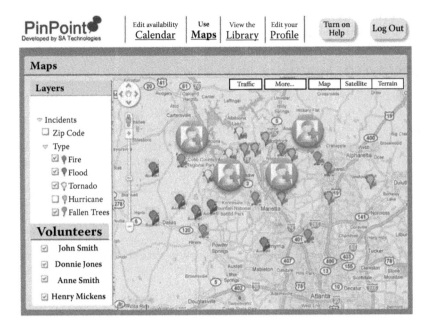

FIGURE 8.5   Illustrative example of PinPoint™ user interface.

CONCLUSIONS

In this chapter, we described two case studies that highlighted cultural ergonomics issues that are relevant in disaster management: (1) meeting the collaboration and coordination needs of CDC biosurveillance teams with diverse background, training, and experience, and (2) meeting the specific human performance needs of Disaster Action Teams composed of volunteers, many of whom are older adults (age 65 and up). We proposed that these challenges can be addressed by applying sound human factors and HCI principles, with a particular emphasis on promoting individual and team situation awareness. Using these two case studies, we demonstrated the application of our SAOD approach for developing disaster management systems with consideration for relevant cultural ergonomics issues.

Our SAOD approach can be applied to generate technology solutions to address other cultural ergonomics issues in complex domains such as disaster management. For example, the GDTA methodology can be utilized to identify the unique SA requirements of individuals with varying educational levels. Users with lower educational levels may require greater scaffolding to support their understanding of task-relevant information. In turn, relevant SAOD Principles can be applied to create useful system interfaces that effectively provide these users with critical information requirements while flexibly adapting to meet their specific information demands (e.g., features and functions that minimize task complexity and avoid placing undue strain on the user's memory). Promoting situation awareness in individuals and teams is an important design objective for any domain where the effects of ever-increasing technological and situational complexity on the human decision-maker are a concern. Our principled SAOD approach is ideally suited to address this design objective.

ACKNOWLEDGMENTS

The opinions, views, and conclusions contained herein are those of the authors and should not be interpreted as representing the official policies, either expressed or implied, of the organizations with which the authors are affiliated. Portions of this chapter were presented at the International Ergonomics Association 17th World Congress on Ergonomics, Beijing, and at the Third International Conference on Applied Human Factors and Ergonomics, Miami.

REFERENCES

AgeLight. 2001. Interface design guidelines for users of all ages. AgeLight.

Bolstad, C. A. 2001. Age-related factors affecting the perception of essential information during complex driving situations. Unpublished Dissertation, North Carolina State University, Raleigh.

Bolstad, C. A., and T. M. Hess. 2000. Situation awareness and aging. In *Situation Awareness Analysis and Measurement*, ed. M. R. Endsley and D. J. Garland, 277–302. Mahwah, NJ: Lawrence Erlbaum.

Bolstad, C. A., H. M. Cuevas, J. Wang-Costello, M. R. Endsley, W. J. Page, and T. Kass-Hout. 2011. Integrating human capabilities into biosurveillance systems: A study of biosurveillance and situation awareness. In *Biosurveillance: Methods and Case Studies*, ed. T. Kass-Hout, and X. Zhang, 79–94. Boca Raton, FL: CRC Press–Taylor & Francis Group.

Centers for Disease Control and Prevention (CDC). 2010. CDC Vision for the 21st Century, http://www.cdc.gov/about/organization/mission.htm (accessed January 22, 2011).

Endsley, M. R. 1995. Toward a theory of situation awareness in dynamic systems. *Human Factors* 37(1): 32–64.

Endsley, M. R., B. Bolte, and D. G. Jones. 2003. *Designing for Situation Awareness: An Approach to Human-Centered Design*. London: Taylor & Francis.

Endsley, M. R., and W. M. Jones. 2001. A model of inter- and intrateam situation awareness: Implications for design, training and measurement. In *New Trends in Cooperative Activities: Understanding System Dynamics in Complex Environments*, ed. M. McNeese, E. Salas and M. R. Endsley, 46–67. Santa Monica, CA: Human Factors and Ergonomics Society.

Fricker, R. D., Jr., B. L. Hegler, and D. A. Dunfee. 2008. Comparing syndromic surveillance detection methods: EARS' versus A CUSUM-based methodology. *Statistics in Medicine* 27 (17): 3407–29.

Garrett, S. K., and B. S. Caldwell. 2009. Human factors aspects of planning and response to pandemic events. *Proceedings of the 2009 Industrial Engineering Research Conference*, May 30–June 3, 2009, in Miami, Florida.

Hawthorn, D. 2000. Possible implications of aging for interface designs. *Interacting with Computers* 12: 507–528.

Hollinworth, N. 2009. Improving computer interaction for older adults. *ACM SIGACCESS Accessibility and Computing*, 11–17. New York: ACM.

Jones, R., M. Mossey, and M. R. Endsley. 2010. Website usability design principles for the elderly. In *Proceedings of the Third International Conference on Applied Human Factors and Ergonomics (AHFE)*, ed. G. Salvendy and W. Karwowski, 833–841, July 17–20, 2010, in Miami, FL. Boca Raton, FL: Taylor & Francis.

Kaplan, M. 1995. The culture at work: Cultural ergonomics. *Ergonomics*, 38, 606–615.

Nielsen, J. 2002, April 28. Usability for Senior Citizens, http://www.useit.com.

Rolka, H., D. Walker, and J. Heitgerd. 2009, June 18. An overview of CDC's OCIIX BioPHusion Program, www2.cdc.gov/ncphi/disss/.../June2009_BioPhusion_Seminar_508.ppt (accessed January 22, 2011).

Strater, L., S. Scielzo, M. Lenox-Tinsley, C. A. Bolstad, H. M. Cuevas, D. M. Ungvarsky, and M. R. Endsley. 2009. Tools to support ad hoc teams. *Proceedings of the Advanced Decision Architectures Final RMB Workshop*, ed. P. McDermott and L. Allender, 359–377, July 22–23, 2009, in Washington, DC.

Wagner, M. M., A. W. Moore, and R. M. Aryel. 2006. *Handbook of Biosurveillance*. St. Louis, MO: Elsevier Science & Technology Books.

Watanabe, M., S. Yonemura, and Y. Asano. 2009. Investigation of web usability based on the dialogue principles. In *Human Centered Design*, ed. M. Kurosu, 825–832. Berlin: Springer-Verlag.

9 Cultural Ergonomics Implications in Forensics

Marc L. Resnick and Tonya Smith-Jackson

CONTENTS

At first glance, the manner in which culture has been scoped in this book does not seem to relate to how researchers and practitioners might assist in providing information to understand and answer legal questions that arise, especially as these legal questions relate to the design, evaluation, and use of systems, processes, and products. As stated in Chapter 1, we have bounded culture using a cultural psychology definition that highlights common history, normative practices, geography, language, beliefs and customs, power arrangements, and kinship as the attributes of culture on which we focus (Veroff and Goldgerger, 1995). These attributes do, in fact, play a role in how individuals process information and interact with systems because they determine the content, representations, and mental models that comprise the complex networks and schemas in long-term memory; hence the labeling of culture as a meta-schema. Understanding culture as a meta-schema helps to frame how we approach forensic analysis of cases and events to identify design issues that might have differential impacts on users from different cultures. Certainly, as engineers, psychologists, and designers across the spectrum, user capabilities, preferences, and experience can vary because of differences in the diffusion of technologies, technology access, and lived experience. While there may be the rare occasion of user attributes that seem to be universal, the assumption that all users bring a common set of beliefs, attitudes, and behaviors to any interaction is a misconception, and one that, if applied to how systems and products are engineered and marketed, might very well lead to outcomes that result in litigation. Misguided or culturally incompatible design can have negative consequences for certain groups. Some examples of how culture influences our work in human factors forensics can be found in such areas as product design, risk perception and allocation of responsibility by jurors.

PRODUCT DESIGN

As new and existing manufacturers design, test, and release products to be used by the general public and by workers in occupational settings, the risk of injuries and fatalities due to defective product design remains a concern among researchers and practitioners in human factors and ergonomics. Specifically, those concerns arise from a seeming lack of value placed on research to understand users' capabilities and limitations and a failure to test prototypes and final product versions in an ecologically valid context before release. Defective product design often includes the product itself as well as the warnings and risk communications associated with the product. Although manufacturers may have a history of conducting reliability and safety testing from a technical perspective (i.e., drop tests, battery function testing, electrical safety testing), they may not conduct valid and reliable usability and safety tests that account for human-system/product interaction within the context of use. Federal government guidance, voluntary standards, and a vast array of textbooks, research articles, and guidelines are available to manufacturers, but even these basic tools tend to be under-utilized. If used, problems associated with general and cultural user attributes might be identified and remedied before products are released.

An example of what is easily accessible in the public domain, the Food and Drug Administration (FDA) released industry guidance on July 18, 2000, that provided information on how to integrate human factors engineering to minimize problems associated with medical device usage (FDA, 2011). In this guidance, the FDA encouraged industry to address use-related hazards "within the context of a thorough understanding of how a device will be used" (p. 5). Interestingly, the FDA, when describing usability evaluation methods such as heuristic evaluation, made reference to the importance of de facto standards, or standards that are not formalized in writing per se, but exist simply because of "social and cultural norms and constraints for the use of device components" (p. 24). Social and cultural norms will determine how devices are adopted and operated within the context of use, whether used in clinical or home environments. A draft guidance with expanded information regarding considerations such as context of use and culture was released on June 22, 2011 (FDA, 2011). Kroemer (2006) reiterates the caveat that there is no normal or common set of user capabilities that are specific to a product or system and describes users as having diverse and varying capabilities and limitations. Only usability and user-centered safety evaluations can unveil the problems that need to be identified. Another example can be found in the ANSI Z535 2007 series that added more comments about the importance of understanding user diversity that is grounded in culture (ANSI, 1996). Cultural attributes should also serve as the user framework for the analysis of products whose use results in severe injuries and fatalities.

Social and cultural norms drive how the product will be used within users' environments (contexts of use). Thus, legal questions arise regarding ordinary and foreseeable use of a product which impose legal responsibilities on manufacturers to produce safe and usable products. From medical products to manufacturing equipment to household cleaning chemicals, the importance of norms will drive how users interact with products, and norms are culturally derived. For instance, severe injuries such as burns and impact traumas involving stove tipping have occurred for many years. Stove tipping can occur because of improper bracing or bracketing of the stove when it is installed (due to poor or misleading instructions) and can also be related to social and cultural norms among specific socioeconomic status groups. For instance, stove tipping can occur when children climb on oven doors that are opened while the oven is in operation. While some might wonder why anyone would leave a stove open for any length of time while the oven is on, most who are knowledgeable of user behaviors that occur within some contexts of poverty might very well foresee this practice. Within and outside of the United States, ovens are often used early in the morning as home heating sources for users who are financially challenged or live in rented apartments that do not have reliable or effective heating. For many users, ovens serve multiple functions beyond cooking, to include heating sources in place of unaffordable space heaters and as a way to store household objects that are either related to cooking or need to be hidden to prevent theft (since burglaries can be common in some neighborhoods). The users' multifunction mental model of ovens and stoves may not match some designers'/manufacturers' mental model. But, manufacturers can conduct usability evaluations with target customers, conduct expert reviews, and do their own archival research to identify these and similar norms and customs that will impact how the product is used. An expert who needs to analyze accidents resulting from stove tipping should consider the extent to which the manufacturer has conducted responsible and thorough product testing and the extent to which user social norms and culture impact how the product is used.

Similar problems occur when manufacturers transfer products initially designed for use in North American cultures to other global settings. At the simplest level, anthropometric dimensions can be significantly different from one location to another. Resnick and Corredor (1995) concluded that industrial equipment designed for 95% of the US workforce would accommodate only 50% of the Colombian population, regardless of the fact that many Colombian manufacturers source their equipment directly from US vendors.

Task procedures also vary between cultures. For example, in the United States, front loading from the torso is a common position to carry objects, and safety and health guidelines for lifting and carrying in manual materials handling operate on that assumption. Manufacturers who apply front-loading guidelines to design packaging or products to minimize risk to the back, may be placing users from other nations at unreasonable risk of harm when the product is shipped to other countries. In many other countries, head and shoulder loading are common methods to carry materials, but this common practice may not be considered in the design of products manufactured in the United States (Datta, Chaterjee, and Roy; 1975; Heglund, Willems, Penta, and Cavagna, 1995; Makin and Ghanem, 1995; Manuel, 2003; Minetti et al., 2006).

In addition to technology transfer problems, people also transfer from one nation to another. User customs applied to work and product use may be different within the same environmental contexts, such as occupational settings. This, too, is a factor that is important in manufacturers' risk management and usability evaluation activities.

Oh, Iridiastadi, and Smith-Jackson (2011) identified several cultural critical incidents that were associated with product use as reported by respondents to a questionnaire (n = 55) who self-identified as U.S. ethnic minorities or from other nations, including Ethiopia, Eritrea, India, Indonesia, Peru, China, and South Korea. Cultural critical incidents (CCIs) were defined as events caused by design conflicts arising from a lack of knowledge of different cultures. Most of the types of CCIs reported were related to safety and usability. This included instances where the products or instructions were confusing or misleading because of design. The contexts in which CCIs occurred were most commonly associated with workplace, educational, and business use or in leisure settings (home or recreation), as opposed to religious, home, or medical settings. Manufacturers should understand their target users, especially as it relates to demographics that might impact culture. These include nationality, socioeconomic status, ethnicity, and gender. Considerations of these cultural attributes do not require any additional steps beyond what manufacturers should do when routinely targeting and recruiting users for participation in product usability and safety evaluations.

Questions also arise when conducting hazard and accident analyses regarding the intended user. Who is the intended user of a product? All manufacturers should understand the intended user of the products they design. Without that knowledge, manufacturers risk releasing a product that is unreasonably dangerous. A lack of knowledge of the intended user also makes it difficult for manufacturers to meet their duty to warn by ensuring they design warnings that are effective for the intended users. Point-of-sale demographics should not be relied on as the sole means to identify users, since any given set of users is larger than the set of purchasers for the same product. As an example, farm owners who speak English as their first language may purchase agricultural equipment and pesticide products, but the actual users may include farmworkers whose first language is Spanish (Hispanic/Latino) or French (Haitian). Linguistic usability also includes more than simply translating material from English to other languages, but it requires further testing to ensure the translations have fidelity in terms of conveying the same mental representations and concepts. Even translations from other languages into English may not have high fidelity for those who speak English as a first language. Similar problems can occur, as demonstrated in the photo shown in Figure 9.1. This photo was taken in Japan and shows a storefront window that is used to communicate laws about wearing pants. In this case, the translation could lead to serious misunderstandings about the consequences of not wearing pants.

Additionally, user capabilities may vary within the same context of use. For instance, professionals such as home health workers or caregivers who are family members might purchase medical products such as wheelchairs or infusion pumps, but additional users of those products will include patients themselves who must also operate the equipment. Any given product might have several user groups, whose cultural and social norms should be understood and accounted for in design and evaluation. Recruitment of users for research participation can be challenging, but it is certainly worth the investment of time and resources in the long term. It is important to ensure the sample recruited for design and evaluation is sufficiently representative of the intended user population. As indicated in the methods discussion

FIGURE 9.1 Translation problems associated with multiple-language warnings. The statement in English reads "Yes! We Pants! No Pants! No LIFE!" It is not clear what the actual laws and codes are meant to be conveyed.

(this volume), designers should ensure the subsample sizes are adequate to identify any problems that occur because of cultural incompatibilities. Smith-Jackson (2006; Smith-Jackson et al., 2011) identified specific strategies to recruit users and conduct culturally-competent testing and evaluation. These included using researchers who match the demographics of the intended users, developing rapport with community and religious organizations to recruit participants, and partnering with minority-serving institutions who can provide faculty with significant expertise and connections to certain communities. With regard to ensuring researchers are compatible with the intended users, the compatibility between the ethnicity of the users and the ethnicity of the researcher (e.g., interviewer, focus group moderator, usability tester) might impact the quality of the data and number of usability problems identified by users; higher compatibility is associated with users reporting a higher number of valid usability problems (Vatrapu and Pérez-Quiñones, 2006).

RISK PERCEPTION

One key feature of users from the perspective of design is how users perceive the risk of a product or system. Risk perception influences behaviors. If subjective risk is lower than objective risk of a product, users are placed at risk of harm because the assumptions and precautionary behaviors they deploy when using the product may not be adequate to protect them from harm. Similarly, if subjective risk is greater than objective risk of a product, users may underutilize the product or use it in incorrect ways which ironically increase the objective risk. There are several factors found to influence risk perception that have implications for how we approach human factors forensics.

Gender, a cultural schema, is one such factor. Gender schemas are established through years of socialization beginning at the time of birth (Eagly, 1983, 1992, 1995; Josselson, 1987; Lott and Maluso, 1993) and continuing through the life span. Gender is understood in the scientific literature to influence behavior and attitudes through socialization, and the extent to which a person is gender-typed is based on the strength of the gender socialization (Bem, 1981; Eagly, 1983, 1992, 1995; Josselson, 1987; Lott and Maluso, 1993). Several studies have identified the role of gender in risk perception, so it is prudent that manufacturers consider these differences. Unfortunately, the scientific treatment of gender in research has often neglected the complexities of analyzing differential impacts of product design on different genders, choosing only to superficially compare men to women (Finucane et al., 2000; Messing et al., 2003). These types of analyses continue to support evidence of disparities in health and safety between genders, but the only conclusions that are drawn are that the outcomes are "different." Few researchers provide substantial analyses of the differences to determine how to provide equivalent benefits to both groups when products are designed. For instance, Vredenburgh and Zackowitz (2006) summarize the effects of gender and ethnicity on risk perception. They describe a study by LaRue and Cohen (1987) that found that women tend to perceive higher levels of risk, and this leads to an increased likelihood to look for and read warnings. This leads women to generate greater expectations of negative outcomes and to take greater precautionary measures, including greater use of personal protective equipment. In one specific example from Vredenburgh and Cohen (1995a), women perceived a higher risk when engaged in recreational activities such as skiing and scuba diving. This leads to a greater likelihood that women would read warnings and comply with them. Smith-Jackson (2006) also suggests that gender differences are based on socialization and societal norms as opposed to physiological or cognitive differences. Socialization leads to differences in beliefs and attitudes toward risk and towards behavior. It also may influence the types of hazards one is exposed to because of societal norms in job assignments, recreational activities, and household responsibilities. She cites contradictory results in how gender impacts the likelihood to become aware of warning signs. Goldhaber and deTurck (1989) describe an example where female students were less likely to be aware of warning signs at a swimming pool, whereas Godfrey, Allender, Laughery, and Smith (1983) found that females were more likely to read and comply with warnings.

Additional studies by O'Brien and Atchison (1998) and Glover and Wogalter (1997) found that females viewed warnings as more credible and were more likely to seek more information and comply with recommended precautions after reading warnings about an earthquake. Differences in attitude towards risk could be mediated by psychological factors such as locus of control and or self-efficacy. Smith-Jackson (2004) found that increased levels of risk perception could be explained by locus of control and self-efficacy. Perhaps females feel less in control of their risk for injury and therefore feel a greater need to take protective measures such as reading or complying with warnings. Self-efficacy could work both ways. The perception that one is less capable of protecting oneself could lead females to be more reliant on protective safety practices. Or this belief could lead to the extension that they will gain less from the protective safety practices and therefore would not gain from applying them. Clearly, additional research is required to resolve these potential implications.

Ethnicity has also been shown to influence risk perception. Vredenburgh and Zackowitz (2006) summarize a study by Vredenburgh and Cohen (1995b) that investigated the differences in risk perception among a variety of ethnic groups living in the United States. Using a survey methodology, the study found that Hispanic, Caucasian, and African-American participants perceived higher levels of risk than Asian participants for a variety of work and recreational activities. Vredenbrugh and Zackowitz speculate that the source of these differences lies in the formation of expectations regarding the potential for injury when engaged in these activities. If populations of different ethnicities that reside in the same country have significantly different perceptions of risk, it is likely that populations residing in different countries will have even greater differences in risk perception.

There is a dearth of research investigating how differences in risk perception affect behavior, specifically with regard to protective safety practices. Data enumerating varying injury rates between ethnic groups or genders would provide insight into how culturally derived differences in risk perception affect behavior. The order of this effect would also reveal insights into risk and liability. If the risk perception affects behavior then research would need to focus on identifying the reasons for differences in risk perception and interventions should focus on mitigating the difference in risk perception. It could be related to differences in socialization, media exposure, or self-efficacy and locus of control. In this case, the risk perception is not justified and interventions should seek to eliminate it. On the other hand, differences in risk perception could result from differences in injury rates experienced by cultural populations, perhaps due to varying job assignments, work practices, or performance expectations. In this case, a more appropriate intervention would modify the work practices to reduce the injury risk rather than its perception.

ALLOCATION OF RESPONSIBILITY

How consumers and other stakeholders perceive the relative responsibility of parties involved in disputes is a critical domain for human factors research. When a user perceives that the provider of a service is fully responsible for his safety, he may be less likely to look for, read, and/or comply with warnings than when dealing with providers with less of a safe reputation. Consider the different perceptions when riding a roller coaster at a Disney theme park compared to a local fairground. The visitor may be less vigilant and careful at the Disney park because of the strong brand image Disney has created of being a safe and family-friendly company. Ironically, this could lead to an increased risk at the Disney parks and behoove Disney to take extra steps when designing risk communications.

Similarly, when a juror perceives that a product user was ultimately responsible for protecting herself and her family from products purchased at the local hardware store, the juror may not award damages in the lawsuit that results from a subsequent injury. Attorneys may want to consider these differences during the jury selection process or when developing a litigation strategy.

If a citizen perceives that the government is responsible for developing and enforcing workplace safety standards, she may be less willing to fault an employer that provides its workers with a hazardous workplace that nevertheless meets all federal codes and regulations. She may be less likely to fault the company in a subsequent lawsuit or may choose which policy makers to vote for based on their positions on these kinds of issues. This has implications for workplace design policies, allocation of resources at the workplace, and in government.

A body of research often referred to as the "attribution" or "allocation" of responsibility investigates these perceptions, including the effects of culture. Participants are presented with the description of an event that led to an injury or damaging incident and the parties involved—usually including the manufacturer, vendor, purchaser, and user. They are asked to allocate the responsibility for the injury or damage among these groups as if they were a member of a jury. Culture can be investigated by varying the cultural demographics of the participants or of the parties described in the scenarios. Darden et al. (1991) discovered that sympathy in product liability litigation depends on the values of the juror, which can vary significantly among cultures. This has been investigated with Asian (Skirloff, 1997) and Hispanic (Mathews, 1996) populations. These studies investigated aspects such as family values, socioeconomic status, and life experience, all of which influenced sympathy for different parties and allocation of responsibility among them. This is not just true for native cultures. Immigrants to the US maintain their original cultures for some time, often by living in ethnic enclaves and associating with individuals from their culture (Miller 1993).

Resnick and Jacko (1998) conducted a study that specifically investigated the attribution of responsibility for safety among different ethnic groups. They compared the attributions of responsibility for injuries to children of Hispanic and Asian populations with a previous study that focused on the native US population (Laughery et al. 1996). They found that in the Asian group children became as responsible as their parents at age 12. In their Hispanic group, children did not achieve this level of responsibility until age 16. This compared to the Laughery et al. (1996) study, which found this crossover at age 13. They labeled this crossover point a potential "coming of age" of children at which they take over responsibility for their own safety and behavior from their parents.

Resnick and Jacko (1998) also found significant differences in the allocation of responsibility to the product manufacturer. The Hispanic cohort attributed 25% of the responsibility to the manufacturer compared to 30% of for the Asian cohort and 40% among the Laughery et al. (1996) US population. The speculated that the lower attribution of responsibility to manufacturers among the ethnic populations could reflect different standards of care or legal requirements in their countries of origin or from inherent cultural differences in perceptions of individual responsibility. It would be very interested to know which of these two was the source of the difference. Perhaps differences in cultural perceptions influence how legal systems and standards develop. Or perhaps differences in experience longitudinally have an impact on the development of cultural expectations and values.

Gunia, Sivanathan, and Galinsky (2009) found that psychological connections between an original and a follow-on decision maker can lead to sunk cost fallacy and escalation of commitment. This may have significant implications for allocation of responsibility. If jurors have such connections with defendants, such as shared cultural attributes or perspectives, they may discount errors made by the defendant in finding guilt or setting penalties. In addition to selecting for specific cultures, litigators may want to match the culture of their client with those of the selected jury.

APPLICATIONS

Tools to conduct design for users who represent diverse cultures are few and far between. We provide a checklist that is based on a literature review by Smith-Jackson and Essuman-Johnson (2011) to support culturally competent research and design. Table 9.1 serves as a checklist that can be used to guide product evaluations

TABLE 9.1
Checklist of Attributes of Culturally Competent Product Research and Development

	Attribute	
1	Usability problems are identified and reported by target users	_____
2	Product addresses real-world needs and requirements of target users	_____
3	Research and development values subjective experiences of target users	_____
4	Research and development values objective measures to complement subjective experiences	_____
5	Research and testing methods mix qualitative and quantitative instruments, procedures	_____
6	Research and testing methods account for linguistic framing and differences in conceptual representation	_____
7	Marketing efforts are truthful and equivalent for various cultural groups	_____
8	Methods can be implemented within the situational context of the target users	_____
9	Methods are implemented by culturally competent researchers or indigenous research assistants	_____
10	Methods are pilot-tested on a subset of target users	_____
11	Methods are as portable or mobile as possible (for hard-to-reach user groups)	_____
12	Data to be analyzed qualitatively can be transcribed into equivalent forms	_____
13	Data to be analyzed quantitatively are checked for distributional equivalence	_____
14	Data to be analyzed quantitatively are checked for reliability equivalence	_____
15	Data to be analyzed are disaggregated by cultural groups	_____
16	Results are translated using relevant cultural frames of reference	_____
17	Results are transformed to reflect the actual data patterns	_____
18	Results are generalized consistent with statistical or hermeneutic frameworks	_____
19	Results are verified by subject matter experts and indigenous representatives	_____
20	Results are shared with the target communities and applied to product design	_____

by examining whether manufacturers (designers, developers) have shown cultural competence when designing and evaluating products that target multiple user groups. The table has been modified from the original Smith-Jackson and Essuman-Johnson checklist.

CASE STUDY

This hypothetical case illustrates how differences in risk perception between ethnicities, genders, and users of varying socioeconomic status can lead to real differences in behavior and consequences for their safety—even when everything else about the situation is the same.

Consider the kerosene lantern illustrated in Figure 9.2. This identical lantern was purchased by three different consumers:

- Jane is a wealthy Caucasian woman. She used the lantern when camping with her daughter's Brownie troop. Because of the nature of the trip, the whole group was very focused on safety. But Jane felt out of her element because she was not an avid camper. In fact, camping really was not "her thing." She felt a little self-conscious because she didn't know much about what is important about safety when using a camping lantern.

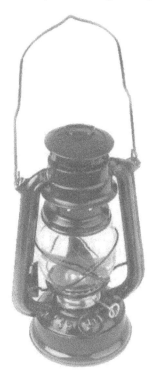

FIGURE 9.2 Camping lantern that is the subject of the end-of-chapter case.

- Lee is a Chinese-American working-class man. He immigrated as a child, but still predominantly uses Chinese as his primary spoken and written language, although he understands English reasonably well. He used the lantern as part of a religious ceremony for his family. During the ceremony, he was very focused on the specific rituals of the ceremony—trying to get all of the nuances correct for the sake of his family's pride. He had a lot of experience using the lantern and similar products from previous ceremonies. His previous successful experiences suggested that he had little to worry about when it came to safety.
- Claude is a survivor of the 2009 earthquake in Haiti and was using the lantern, donated to him by the Red Cross, while sheltered in a makeshift tent on the outskirts of Port au Prince. He was cramped into a tent city with thousands of other earthquake victims. He was concerned about an open flame in such a flammable environment, but he had no choice because it was all he had available. The kerosene fuel he was using was something he had scrounged when the original supply ran out. He was not sure how this would affect the working of the lantern, but again he had no other options.

In all of these cases, the individual is using the lantern in a family-sized tent/shelter with his or her spouse and two children ages eight and fourteen inside. In all cases it is 8 pm, and the sun set two hours previously. The weather and ground are both reasonably dry. The same amount of clutter is present inside the tent. Everything is basically the same in all three cases except the cultural and context described above. In all three cases, the warning text shown in Figure 9.3 was present in exactly the same place on rear of the lantern.

Now let's consider each of the four main steps in the risk communication process: noticing the warning, reading the warning, comprehending the warning, and complying with the warning. How would they be different for members of these three very different cultures?

WARNING

As with all devices that use a flame and flammable gas, safety is first and foremost. There are several safety recommendations that must be adhered to in order to safely enjoy this lantern, including:

- Lantern top and globe will become very hot and can cause serious burns. You need to be especially careful around children because the lanterns are very hot.
- Fuel is flammable and should be handled carefully. Keep fuel out of reach of children and away from heat. When burning your lantern, set it on a flat, stable surface, so it does not tip over and catch anything on fire.
- Only adults should use fuel-based camping lanterns.
- Presoak new wicks in fuel before installing them in the lantern.
- Always follow the manufacturer's directions.

FIGURE 9.3 Warning text located on the rear of the lantern.

NOTICED AND READ

Whether or not a user notices a warning generally depends on two factors, the salience of the warning and whether the user is proactively motivated to look for a warning. In all three of these cases, the salience is the same, so it is the proactive seeking that we need to consider.

- Jane has no problem with English or reading at an eighth grade level. Jane is intentionally focusing on safety, so she looks for any safety information that might be present on the lantern. She has no trouble finding the warning label attached to the back side, even in the dark of the night. She not only reads it, but tries to explain it to the girls as part of the exercise.
- Lee's past success gives him little reason to be proactive. He is very focused on the ritual activities of the ceremony, so he never looks at the back of the lantern. He doesn't even notice that the warning is present. Even if he did, the ritual is pretty specific so he doubts that he would change the way he used the lantern regardless of what it says. Culturally, the importance of the religious procedures takes precedence over safety considerations. As a male, his high self-efficacy makes him more confident that he can handle anything that comes up.
- Claude is very motivated to use the lantern because it is essential to his family's happiness. He is partially focused on safety because of the alternative fuel he is using. So he would probably notice the warning at some point during this extended use. If he noticed it, he would at least scan it to see if there was anything about the fuel.

COMPREHENSION

Comprehension has two parts. First, the user has to understand the language in which the text is written. This warning, as with most risk communication text, is written at an eighth-grade reading level, and as an American product is written in English. Even when exported to other countries, warnings are often not translated into the home language (in the case of the third example, Creole), and immigrants are rarely supported in their native language. Second, the user has to understand the meaning of the warning. What are they supposed to do and why is this important?

- Jane has no problem either with the English or with the eighth-grade reading level. She understands all of the words and can accurately read them to the girls in the troop. But her unfamiliarity with camping has her uncertain about what some of the terms are referring to. She can't recognize whether the fuel is correct, whether it is too close to the tent wall, or how old the girls have to be to be considered "adult" according to the warning. Resnick and Jacko (1998) suggests that she will consider the 14-year-old sufficiently responsible to handle the lantern, but not the 8-year-old.
- Lee is a well-educated user and has no trouble reading or understanding the warning, even in English. He has used the lantern many times in the past, and knows what each term refers to. His ceremony does not violate any of the warning proscriptions.

- Claude speaks only French. Some of the words are similar enough that he can get a basic sense, but really doesn't understand what it is telling him to do, or why. There is no research on the age at which the Haitian population considers children to be responsible for risky activities.

COMPLIANCE

Warnings are not particularly helpful if they are ignored. They need to convey enough likelihood and severity to convince the user that the recommended behavior is worth the effort. Compliance is often the place where risk communication has its greatest challenge. We see that here in three ways.

- Jane is highly motivated to model good behavior to the girls in the troop by following everything in the warning. Her socioeconomic status reinforces this motivation. But she isn't quite sure how. As a female, she may have a lower self-efficacy that could increase this uncertainty or could lead her to be more careful about compliance.
- Lee is not able to comply with anything in the warning that violates the religious ritual because that is the most important thing to him at the moment.
- Claude has difficulty putting any emphasis on safety because of his situation and socioeconomic status. He is pretty much stuck making do with whatever he has available. He can try to be careful, but he would take basic precautions whether there was a warning or not. His gender and socioeconomic status increase his self-attribution of responsibility and how much his family will look to him to establish safety practices.

REFERENCES

Bem, S. L. (1981). Gender schema theory: A cognitive account of sex typing. *Psychological Review,* 88, 354–365.

Darden, W. R., DeConinck, J. B., Babia, B. J., and Griffin, M. (1991) Consumer sympathy in product liability suits. *Journal of Business Research,* 22, 1, 65–89.

Datta, S. R., Chaterjee, B. B., and Roy, B. N. (1975). Maximum permissible weight to be carried on the head by a male worker from eastern India. *Journal of Applied Physiology,* 38, 132–135.

Eagly, A. H. (1983). Gender and social influence: A social psychological analysis. *American Psychologist,* 38, 971–981.

Eagly, A. H. (1992). Uneven progress: Social psychology and the study of attitudes. *Journal of Personality and Social Psychology,* 63, 693–710.

Eagly, A. H. (1995). The science and politics of comparing women and men. *American Psychologist,* 50, 145–149.

Finucane, M., Slovic, P., Mertz, C., Flynn, J., Satterfield, T. (2000). *Health, Risk, and Society,* 2, 159–172.

Food and Drug Administration (2000). *Guidance for Industry and FDA Premarket and Design Control Reviewers. Medical Device Use-Safety: Incorporating Human Factors Engineering into Risk Management.* Department of Health and Human Services, Food and Drug Administration, Center for Devices and Radiological Health. Online: http://www.fda.gov/downloads/MedicalDevices/DeviceRegulationandGuidance/GuidanceDocuments/ucm094461.pdf. Retrieved December 5, 2011.

Food and Drug Administration (2011). *Draft Guidance for Industry and Food and Drug Administration Staff: Applying Human Factors and Usability Engineering to Optimize Medical Device Design.* U.S. Department of Health and Human Services, Food and Drug Administration, Center for Devices and Radiological Health. Online: http://www.fda.gov/downloads/MedicalDevices/DeviceRegulationandGuidance/GuidanceDocuments/UCM259760.pdf. Retrieved December 5, 2011.

Glover, B. L., and Wogalter, M. S. (1997). Using a computer simulated world to study behavioral compliance with warnings: Effects of salience and gender. In *Proceedings of the 41st Annual Meeting of the Human Factors Society*, Santa Monica, CA: Human Factors and Ergonomics Society. pp. 1283–1287.

Godfrey, S. S., Allender, I., Laughery, K. R., and Smith, V. L. (1983) Warning messages: will the consumer bother to look. In *Proceedings of the 27th Annual Meeting of the Human Factors Society*. Santa Monica, CA: Human Factors and Ergonomics Society. pp. 950–954.

Goldhaber, G. M. and deTurck, M. A. (1989). Effectiveness of warning signs: Gender and familiarity effects. *Journal of Products Liability*, 11, 271–284.

Gunia, B. C., Sivanathan, N., and Galinsky, A. D. (2009). Vicarious entrapment: Your sunk costs, my escalation of commitment. *Journal of Experimental Social Psychology.* 45, 1238–1244.

Heglund, N. C., Willems, P. A., Penta, M., and Cavagna, G. A. (1995). Energy saving gait mechanics with head-supported loads. *Nature*, 375, 52–54.

Josselson, R. (1987). *Finding Herself: Pathways to Identify Development in Women.* San Francisco: Jossey-Bass.

Kroemer, K. H. E. (2006). "Extra-ordinary" Ergonomics. HFES Issues in Human Factors and Ergonomics Series, 4. Santa Monica, CA: Taylor and Francis.

Laughery, K. R., Lovvoll, D. R., and McQuilkin, M. L. (1996). Allocation of responsibility for child safety. *Proceedings of the Human Factors and Ergonomics Society 40th Annual Meeting.* Santa Monica, CA: Human Factors and Ergonomics Society.

LaRue, C., and Cohen, H. H. (1987). Factors affecting consumers' perceptions of product warnings: An examination of the differences between male and female consumers. In *Proceedings of the 31st Annual Meeting of the Human Factors Society*. Santa Monica, CA: Human Factors and Ergonomics Society. pp. 610–614.

Lott, B., and Maluso, D. (1993). The social learning of gender. In A. E. Beall and J. Sternberg (Eds), *The Psychology of Gender*. New York: Guilford Press. pp. 99–126.

Makin, M. and Ghanem, J. (1995). Reduction of dorso-lumbar angulation by head loading. *Gait and Posture,* 3, 66–71.

Manuel, J. (2003). The quest for fire: hazards of a daily struggle. *Environmental Health Perspectives,* 111, 28–33.

Mathews, R. (1996). Knowing your customers: Serving the needs of your customers takes on a whole new merchandising twist when it comes to emerging minority groups, like Mexican/Latinos. *Progressive Grocer*, 75, 5, 85–87.

Messing, K., Punnett, L., Bond, M., Alexanderson, K., Pyle, J., Zahm, S., Wegman, D., Stock, S., DeGrosbois, S. (2003). Be the fairest of them all: Challenges and recommendations for the treatment of gender in occupational health research. *American Journal of Industrial Medicine,* 43, 618–629.

Miller, C. (1993). Researcher says U.S. is more of a bowl than a pot. *Marketing News*, 27, 10, 6–7.

Minetti, A., Formenti, F., and Ardigo, L. (2006). Himalayn porter's specialization: Metabolic power, economy, efficiency and skill. *Proceedings of the Royal Society B,* 273, 2791–2797.

National Electrical Manufacturers Association (2006). *American National Standards Institute Safety Signs and Colors Standard.* Rosslyn, VA: National Electrical Manufacturers Association.

O'Brien, P. W., and Atchison, P. (1998). Gender differentiation and aftershock warning response. In E Enarson and B. Morrow (Eds), *The Gendered Terrain of Disaster: Through Women's Eyes.* Westport, CT: Greenwood. 173–180.

Oh, C., Iridiastadi, H., and Smith-Jackson, T. (2011). Cultural critical incidents in design: An international perspective. In M. Gobel, C. Christie, S. Zschernack, A. Todd, and M. Mattison (Eds.), *Human Factors in Organisational Design and Management–X, Volume 1*, Grahamstown, Republic of South Africa: Rhodes University, pp. 381–386.

Resnick, M. L. and Corredor, O. (1995). Estimating the anthropometry of international populations using the scaling estimation method. *Proceedings of the 39th Annual Meeting of the Human Factors Society.* Santa Monica, CA: Human Factors and Ergonomics Society.

Resnick, M. L., and Jacko, J. A. (1998). The effect of culture in the attribution of responsibility for accidents involving children in the USA. *Technology, Law, and Insurance, 3, 2,* 173–178.

Skirloff, L. (1997). Generasian gap. *Brandweek*, 38, 26, 20–22.

Smith-Jackson, T. L. (2004). Cultural Ergonomics Education and Training: Evaluation of a Learning Module. *Proceedings of the International Society for Occupational Ergonomics and Safety, Houston, Texas, May 20–24,* 2004, pp. 101–105.

Smith-Jackson, T. L. (2006a). Receiver characteristics. In M. S. Wogalter (Ed.), *Handbook of Warnings.* Mahway, NJ: Lawrence Erlbaum Associates. pp. 335.

Smith-Jackson, T. L. (2006b). Culture and warnings. In M. S. Wogalter (Ed.), *Handbook of Warnings.* Mahwah, NJ: Lawrence Erlbaum. pp. 363–372.

Smith-Jackson, T., and Essuman-Johnson, A. (2011). Culturally-competent ergonomics: A preliminary checklist. In M. Gobel, C. Christie, S. Zschernack, A. Todd, and M. Mattison (Eds.), *Human Factors in Organisational Design and Management–X, Volume 1*, Grahamstown, Republic of South Africa: Rhodes University, pp. 387–392.

Vatrapu, R., and Pérez-Quiñones, M. (2006). Culture and international usability testing: The effects of culture in structured interviews. *Journal of Usability Studies, 1. 4,* 156–170.

Veroff, J., and Goldberger, N. (1995). What's in a name? In N. R. Goldberger and J. B. Veroff (Eds.), *The Culture and Psychology Reader.* New York: New York University Press, pp. 3–21.

Vredenburgh, A. G., and Cohen, H. H. (1995a). High risk recreational activities: Skiing and scuba—What predicts compliance with warnings? *International Journal of Industrial Ergonomics,* 15, 123–128.

Vredenburgh, A. G., and Cohen, H. H. (1995b). Does culture affect risk perception? Comparisons among Mexicans, African-Americans, Asians, and Caucasians. In *Proceedings of the 39th Annual Meeting of the Human Factors and Ergonomics Society.* Santa Monica, CA: Human Factors and Ergonomics Society. pp. 1015–1019.

Vredenburgh, A. G., and Zackowitz, I. B. (2006). Expectations. In M. S. Wogalter (Ed.), *Handbook of Warnings.* Mahway, NJ: Lawrence Erlbaum Associates.

Youn, S. (2005). Teenagers' perceptions of online privacy and coping behaviors: A risk-benefit appraisal approach. *Journal of Broadcasting and Electronic Media,* 49, 1, 86.

10 Future Directions
A Preliminary Research Agenda

Tonya Smith-Jackson, Marc L. Resnick,
and Kayenda Johnson

The intent of this book was to provide a research- and design-based perspective on cultural ergonomics. Our goal was to provide readers with opportunities to explore cultural ergonomics, using a tool kit of methods and skills gained from the application-centered coverage of topics within each chapter. While the role of theory is fundamental to the practice of cultural ergonomics, coverage of theory within one volume is not possible. We have selected an engineering-based approach, providing theory as foundation, but moving quickly to methods and applications. The domain of cultural ergonomics has existed without enough discussion of methods and applications; thus, we hoped to meet this need within the knowledge domain.

Cultural ergonomics continues to evolve as a specialty area, but to advance how we design and conduct research, cultural ergonomics will have to be diffused and integrated throughout all areas of research and design. If cultural ergonomics remains separate and stigmatized, researchers and designers will not internalize the value of conducting inclusive research. To integrate, it is important to develop a research agenda. In this final chapter, we offer a brief list of topical areas to serve as a guide for the development of a research agenda.

Goal 1: Identifying critical cultural factors that should be given priority in the design and evaluation of systems. It is not likely that all cultural differences are necessarily impactful in design and evaluation. More research is needed to develop an ontology to organize and prioritize variables that are crucial in specific engineering contexts relevant to design and evaluation. This should be followed by applying advanced statistical methods and modeling techniques to conduct sensitive and valid analyses of culture.

Goal 2: Understanding the impacts of anthropometric differences on the design of equipment such as personal protective equipment and the extent to which national-ethnic-physical differences undermine the safety and effectiveness of equipment.

Goal 3: Exploration of the cultural embeddedness of interaction metaphors and providing design solutions for inclusive metaphors or for functionality that adapts to users' cultural world views and frames of reference.

Goal 4: Developing methods to conduct forensic analysis to account for the extent to which cultural incompatibility may have contributed to system safety and effectiveness. Additionally, the context of operations of products and systems may be problematic. Product designers may assume one context of use for, for example, a laundry detergent. In the United States, the detergent is likely to be stored in an indoor laundry room. In some rural areas in the United States and in developing countries, the laundry detergent might be used and stored on a back porch or outside near other substances or near a fire. These considerations are important when designing products.

Goal 5: Conducting comparative and sequential studies to identify how culture influences specific variables in specific contexts. Contextual analysis across several cross-cultural scenarios would help to advance the practice of design and would advance research methods and analysis.

Goal 6: Exploring the nature of international collaborations and using macroergonomics methods to identify new variables emerging from cross-cultural collaborations.

Goal 7: Expanding human–systems integration (HSI) models to support the design and evaluation of complex, dynamic systems that will be operated by cross-cultural teams. HSI is in need of an expansion of the view of "people" within the system. Frameworks should be revised to support discourse and inclusive research around the interdependencies specific to the culture of users who are coupled with the system. Research capabilities must focus on disaggregating human factors in order to analyze at the cultural level to support multilevel optimization that produces the same or equivalent benefits and does not produce greater risks to one group compared to another.

Goal 8: Empirically identifying cultural variables that influence the effectiveness and validity of learning and training systems. Most organizations and institutions continue to use outdated teaching-learning models that were replicas of European training and education systems. Ethnic, class, gender, and generational diversity have introduced new demands for more inclusive education and training systems. There are vast opportunities to learn from other cultures who have used certain pedagogies successfully. These include minority-serving institutions in the United States, institutions in developing countries, nomadic educational systems, and local village educational systems.

Goal 9: Developing pedagogies to prepare tomorrow's ergonomists for complex, global systems that will be deployed in cross-cultural contexts. A program to prepare ergonomists, engineers, psychologists, and others who design and evaluate systems is sorely needed, and would require an integration of global concepts, cultural immersion experiences, and education in cultural anthropology. Very strong quantitative and qualitative research training is also required.

Goal 10: Restructuring traditions in biomechanics to address the variety of work and materials handling behaviors that introduce important differences both cross-nationally and within nations (particularly when new immigrants begin working in new nations). Some factors include carrying postures that introduce different lifting and loading demands, such as head loads, shoulder loads, and back loads. Additionally, biomechanics research should use more inclusive contexts for materials handling, which includes carrying loads over rough terrain and without foot protection, and working under conditions of fatigue and strain complicated by nutritional limitations, pregnancy, and conditions such as HIV/AIDS.

Goal 11: Expanding the concept we call "work." Work is done inside and outside of formal organizations. Work includes tasks for everyday living, and many cultures conduct significant amounts of work "off-the-job." These situations include individuals in low-socioeconomic status groups who work multiple jobs, and who may, for example, leave a demanding manufacturing job and immediately work another shift at a services job. Caregivers may complete a "formal" job, only to return home to face additional physical demands of lifting, moving, and bathing children or older family members. In rural areas, individuals face demanding tasks associated with farming and livestock, yet may also work full time in another physically demanding formal organization. Work also occurs in sports settings, which tend to be highly diverse. Paid athletes are undergoing rigorous physical demands while wearing or using poorly designed equipment (i.e., US football helmets, catching mitts and gloves).

Goal 12: Rethinking safety culture and safety climate. Beyond the confusion surrounding these two constructs, as yet no cross-national and inclusive instruments exist to measure them. Cross-national scales have been adjusted and retrofitted, but no global efforts have been implemented to use a psychometric development process that begins with an inclusive sample of participants. This effort needs special attention.

Goal 13: Using inclusive research to advance knowledge domains that have not, as yet, demonstrated a significant focus on the application of culture to design and evaluation. These include transportation, healthcare, law and criminal justice, service industries, and finance and banking.

Index

Milton Keynes UK
Ingram Content Group UK Ltd.
UKHW040102071024
449327UK00019B/750